Lecture Notes in Computer Sc

Commenced Publication in 1973
Founding and Former Series Editors:
Gerhard Goos, Juris Hartmanis, and Jan van Leeuwen

Sargur N. Srihari   Katrin Franke (Eds.)

Computational Forensics

Second International Workshop, IWCF 2008
Washington, DC, USA, August 7-8, 2008
Proceedings

 Springer

Volume Editors

Sargur N. Srihari
State University of New York, University at Buffalo
Center of Excellence for Document Analysis and Recognition (CEDAR)
UB Commons, 520 Lee Entrance, #202, Amherst, New York, 14228-2583, USA
E-mail: srihari@cedar.buffalo.edu

Katrin Franke
Gjovik University College, Norwegian Information Security Laboratory (NISlab)
Teknologivegen 22, 2802 Gjovik, Norway
E-mail: kyfranke@hig.no

Library of Congress Control Number: Applied for

CR Subject Classification (1998): I.5, I.4, I.7.5, I.2.7

LNCS Sublibrary: SL 6 – Image Processing, Computer Vision, Pattern Recognition,
and Graphics

ISSN	0302-9743
ISBN-10	3-540-85302-2 Springer Berlin Heidelberg New York
ISBN-13	978-3-540-85302-2 Springer Berlin Heidelberg New York

Springer is a part of Springer Science+Business Media

springer.com

© Springer-Verlag Berlin Heidelberg 2008
Printed in Germany

Typesetting: Camera-ready by author, data conversion by Scientific Publishing Services, Chennai, India
Printed on acid-free paper   SPIN: 12464401   06/3180   5 4 3 2 1 0

Preface

This Lecture Notes in Computer Science (LNCS) volume contains the papers presented at the Second International Workshop on Computational Forensics (IWCF 2008), held August 7–8, 2008. It was a great honor for the organizers to host this scientific event at the renowned National Academy of Sciences: Keck Center in Washington, DC, USA.

Computational Forensics is an emerging research domain focusing on the investigation of forensic problems using computational methods. Its primary goal is the discovery and advancement of forensic knowledge involving modeling, computer simulation, and computer-based analysis and recognition in studying and solving forensic problems.

The Computational Forensics workshop series is intended as a forum for researchers and practitioners in all areas of computational and forensic sciences. This forum discusses current challenges in computer-assisted forensic investigations and presents recent progress and advances.

IWCF addresses a broad spectrum of forensic disciplines that use computer tools for criminal investigation. This year's edition covers presentations on computational methods for individuality studies, computer-based 3D processing and analysis of skulls and human bodies, shoe print preprocessing and analysis, natural language analysis and information retrieval to support law enforcement, analysis and group visualization of speech recordings, scanner and print device forensics, and computer-based questioned document and signature analysis.

In total, 39 papers from 13 countries were submitted to IWCF 2008, of which 19 (48%) were accepted. We appreciated the number of papers submitted as well as the diversity of the topics covered. We regret that not all manuscripts could be accepted for publication. The review process was a delicate and challenging task for the Program Committee. All manuscripts were carefully reviewed by three experts. We are especially grateful to the 19 members of the Program Committee for their dedication, adherence to high standards, and timely delivery of the reviews.

The organization of such an event is not possible without the effort and the enthusiasm of the people involved. We thank all the members of the Local Organizing Committee. Our special thanks go to Eugenia H. Smith for coordinating the entire organization of the event.

June 2008

Sargur N. Srihari
Katrin Franke

IWCF 2008 Organization

IWCF 2008 was jointly organized by the Center of Excellence for Document Analysis and Recognition (CEDAR), University at Buffalo, State University of New York, USA and the Norwegian Information Security Laboratory (NISlab), Gjøvik University College, Norway.

Workshop Co-chairs

Sargur N. Srihari   University at Buffalo, USA
Katrin Franke   Gjøvik University College, Norway

Local Organization

Eugenia H. Smith   University at Buffalo, USA

Program Committee

Gonzalo Álvarez Marañón   Consejo Superior de Investigaciones Científicas, Spain

Faouzi Alaya Cheikh   Gjøvik University College, Norway
Oscar Cordón   European Centre for Soft Computing, Spain
Edward J. Delp   Purdue University, USA
Patrick De Smet   FOD Justitie, Belgium
Andrzej Drygajlo   Swiss Federal Institute of Technology Lausanne, Switzerland

Cinthia Freitas   Pontifical Catholic University of Parana, Brazil
Zeno Geradts   Netherlands Forensic Institute, The Netherlands

Lawrence Hornak   West Virginia University, USA
Mario Köppen   Kyushu Institute of Technology, Japan
Deborah Leben   US Secret Service, USA
Zhenan Sun   Chinese Academy of Sciences, China
Milan Milosavljević   University of Belgrade, Serbia
Slobodan Petrović   Gjøvik University College, Norway
Hiroshi Sako   Hitachi Central Research Laboratory, Japan
Reva Schwartz   US Secret Service, USA
Cor J. Veenman   University of Amsterdan, The Netherlands
Svetlana Yanushkevich   University of Calgary, Canada
André Årnes   Oracle Norge AS, Norway

Table of Contents

Linguistics

Decision Making and Search

Speech Analysis

Signatures and Handwriting

Computational Forensics: An Overview

Katrin Franke[1] and Sargur N. Srihari[2]

[1] Norwegian Information Security Laboratory, Gjøvik University College, Norway
[2] CEDAR, University at Buffalo, State University of New York, USA
kyfranke@ieee.org,
srihari@cedar.buffalo.edu

Abstract. Cognitive abilities of human expertise modeled using computational methods offer several new possibilities for the forensic sciences. They include three areas: providing tools for use by the forensic examiner, establishing a scientific basis for the expertise, and providing an alternate opinion on a case. This paper gives a brief overview of computational forensics with a focus on those disciplines that involve pattern evidence.

Keywords: Computational science, Forensic science, Computer science, Artificial intelligence, Law enforcement, Investigation services.

1   Introduction

The term "computational" has been associated with several disciplines of human expertise. Examples are computational vision, computational linguistics, computational chemistry, computational advertising, etc. Analogously a body of knowledge and methods to be collectively defined as computational forensics can be defined.

Computational methods find a place in the forensic sciences in three ways. First, they provide tools for the human examiner to better analyze evidence by overcoming limitations of human cognitive ability– thus they can support the forensic examiner in his / her daily casework. Secondly they can be used to provide the scientific basis for a forensic discipline or procedure by providing for the analysis of large volumes of data which are not humanly possible. Thirdly they can ultimately be used to represent human expert knowledge and for implementing recognition and reasoning abilities in machines. While the goal of a computer to provide an opinion is a goal analogous to other grand challenges of artificial intelligence, they are unlikely to replace the human examiner in the foreseeeable future. On the other hand it is more likely that modern crime investigation will profit from the hybrid-intelligence of humans and machines.

More broadly, computer methods and algorithms enable the forensic practitioner to:

- reveal and improve traces evidence for further investigation,
- analyze and identify evidence in an objective and reproducible manner,

S.N. Srihari and K. Franke (Eds.): IWCF 2008, LNCS 5158, pp. 1–10, 2008.

- assess the quality of an examination method,
- report and standardize investigative procedures,
- search large volumes of data efficiently,
- visualize and document the results of analysis,
- assist in the interpretation of results and their argumentation,
- reveal previously unknown patterns / links, to derive new rules and contribute to the generation of new knowledge.

The objective of this paper is to lay the foundations and to encourage further discussions on the development of computational methods for forensic investigation services. Researchers and practitioners in computer science are introduced to specialized areas and procedures applied in forensic casework. Current forensic challenges that demand the development of next-generation equipment and tools are exposed. The forensic scientist and practitioner, on the other hand, are provided with an overview of fundamental techniques available in the computing sciences. Selected examples of successfully implemented computing approaches will help to gain trust in methods and technologies unknown thus far. These examples may also inspire / reveal further forensic areas that can be supported by computer systems.

The remainder of this paper is structured as follows: forensic sciences are briefly described in Section 2. Section 3 aims to provide a definition of computational forensics. In Section 4 the relevant areas of computational / machine intelligence are summarized. Some previous and ongoing studies on computational forensic are provided in Section 5. Section 6 concludes with discussions and points to further directions.

2   Forensics

Forensic science is the methodological correct application of a broad spectrum of scientific disciplines to answer questions significant to the legal system [1]. Technology, methodology and application constitute forensic science and drive its advancement equally. A graph proposed by van der Steen et al. [1] visualizes this interrelation (compare Figure 1). Disciplines involved in forensic sciences are widespread, e.g., biology, chemistry, physics and medicine, and more specialized pathology, anthropology, ballistics to mention a few. With the evolvement of criminal activities, further disciplines are getting involved as for example computer science, engineering and economics. One proposal for categorizing these disciplines in their contribution to forensics is given by Saks [2], who distinguishes

- classical forensic *identification* sciences based on *individualization* (to identify a finger, a writer, a weapon, a shoe that left the mark), and
- more practical-oriented disciplines based on *classification* and *quantization* (chemical, biological, medical, or physical methods) like forensic toxicology.

Forensic sciences use multi-disciplinary approaches to:

- investigate and to reconstruct a crime scene or a scene of an accident,
- collect and analyze trace evidence found,

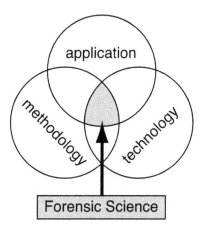

Fig. 1. Forensic science can be defined as the cross section of technology, methodology and application [1]

- identify / classify / quantify / individualize persons, objects, processes,
- establish linkages / associations and reconstructions, and
- use those findings in the prosecution or the defense in a court of law.

The more practical work process of an examination can be summarized as: crime-scene investigation (CSI); documentation / photographing of the scene; questioning witnesses; identification / collection and preservation of evidence; analysis of evidence (e.g. in the laboratory); data integration; link analysis; crime-scene reconstruction; report writing and presentation of findings in court. While forensics has mostly dealt with previously committed crime, greater focus is now being placed on analyzing data gathered to prevent future crime and terrorism [3].

Forensic experts study a broad area of objects, substances (blood, body fluids, drugs), chemicals (paint, fibers, explosives, toxins), tissue traces (hair, skin), impression evidence (shoe or finger print, tool or bite marks), electronic data and devices (network traffic, e-mail, images). Some further objects to be studied are fire debris, vehicles, questioned documents, physiological and behavioral patterns.

Forensic sciences face a number of challenges and demands that are summarized in Table 1. For example they are challenged by the fact that only *tiny pieces of evidence* are hidden in a mostly *chaotic environment*. Examples are a smudged fingerprint on a glass, a half ear print on a door, a disguised handwriting or an unobtrusive paint scratch. The majority of criminals invest all their knowledge and expertise to cover their activities and potential results. Traces have to be studied to reveal *specific properties* that allow for example to identify a person or to link a tool to a caused damage. Moreover, traces found will be *never identical* to known specimen in a reference base, even if traces are caused by the identical source. For example, producing exactly the same tool mark is impossible and printing exactly the same document is impossible. As a

Table 1. Challenges and Demands in Forensic Science

Challenges	Demands
tiny pieces of evidence	sufficient quality of trace evidence
chaotic environment	objective measurement / analysis
specific properties (abnormalities)	robustness & reproducibility
never identical traces	secure against falsification
partial knowledge, approximation	
uncertainties & conjectures	

consequence, reasoning and deduction have to be performed on the basis of *partial knowledge, approximations, uncertainties* and *conjectures*.

In addition to human forensic expertise, the investigative procedure and employed technology decide case resolution. A forensic expert compares traces of evidence on the basis of well-defined sets of characteristics that are primarily based upon domain knowledge and personal experience. Despite great efforts to provide adequate expert training, some forensic methodologies have frequently been criticized, in particular the lack of studies on validity and reliability [2,4]. Attempts have been made to support traditional methods with semi-automatic and interactive systems on the basis of measurements and decisions that lack objectivity and verifiability. Although promising research has been done, computer-based trace analysis is rarely applied in daily forensic casework. Rare exceptions are the fields of digital / computer forensics that use computational methods intrinsically, DNA analysis that takes advantage of algorithms originating from bioinformatics, and databases (e.g. for paint or fine arts), which use mainly manually entered meta information (verbatim) and keywords for data retrieval instead of realistic object presentations and from that derived machine-processable characteristics. Thus necessitating a study of whether forensic sciences can benefit from recent technological developments.

3   Computational Forensics

Computational Forensics (CF) is an emerging interdisciplinary research domain. It is understood as the hypothesis-driven investigation of a specific forensic problem using computers, with the primary goal of discovery and advancement of forensic knowledge. CF works towards

- in-depth understanding of a forensic discipline,
- evaluation of a particular scientific method basis, and
- systematic approach to forensic sciences by applying techniques of computer science, applied mathematics and statistics.

It involves modeling and computer simulation (synthesis) and / or computer-based analysis and recognition in studying and solving forensic problems.

Several terms are currently used to denote mathematical and computing approaches in forensics. *Forensic Statistics* and *Forensic Information Technology* have the longest tradition, yet they are specific. The terms *Forensic Intelligence* and *CF* cover a broader spectrum. It is necessary to establish a sound conceptual framework for CF as in the case of computational vision, computational science, computational medicine, computational biology, etc. The term CF is preferred as it indicates formalization of the methods used by humans, analogous to the use of the term computational vision used by researchers trying to understand biological vision [5]. In this definition computational vision is an attempt to model the visual process by an information processing model. Such a model involves three components: i) a computational theory, ii) methods for representing data and specification of algorithms to process the data, and iii) realization of the algorithms in software and hardware.

A systematic approach to CF ensures a comprehensive research, development, and investigation process that remains focused on the needs of the forensic problem. The process typically includes the following phases:

- analysis of the forensic problem and identifying the goals of study (alternate hypotheses),
- determination of required / given preconditions and data,
- data collection and / or generation,
- design of experiments,
- study / selection of existing computational methods and / or adaptation / design of new algorithms on demand,
- implementation of the experiment including machine learning and training procedures with known data samples, and
- evaluation of the experiment as well as test of the hypotheses.

CF requires joint efforts by forensic and computational scientists with benefits to both. Regarding sharing of knowledge among forensic and computer experts, while there may be good reasons for protecting forensic expertise within a closed community, it would conflict with Daubert and other legal rulings [6], which require the investigative method as being generally accepted, having a scientific basis, etc. The relatively small community of forensic experts can hardly foster scientific bases for their methods independently. As has been successful in the traditional forensic domains (e.g. medicine, biology and chemistry) close cooperation between forensic scientists and computational scientists are possible. In the computational sciences, successful collaborations between computer scientists and biologists, chemists and linguists are known. With these precedents, forensics can benefit from knowledge, techniques and research findings in applied mathematics and computer science. Moreover, several forensic fields cover similar work procedures and tackle similar problems although their investigation objects are different. By means of shared knowledge, sophisticated computational methods can be efficiently adapted to a new problem domain.

The expected impact of CF is potentially far reaching. Most obvious contributions to the forensic domain are to:

- increase *efficiency* and *effectiveness* in risk analysis, crime prevention, investigation, prosecution and the enforcement of law, and to support standardized *reporting* on investigation results and deductions.
- perform *testing* that is often very time consuming. By means of systematic empirical testing scientific foundations can be established. Theories can be implemented and become testable on a larger scale of data. Subsequently, method can be analyzed regarding their strengths / weaknesses and a potential *error rate* can be determined.
- gather, manage and extrapolate data, and to synthesize new *data sets* on demand. In forensics, unequally distributed data sets exist; there are many correct but only a few counterfeit samples. Computer models can help to synthesize data and even simulate meaningful influences / variations.
- establish and to implement *standards* for work procedures and to journal processes (semi)-automatically. Technical equipment also supports the establishment / maintaining of conceptional frameworks and terminologies used. In consequence, data exchange and the interoperability of systems become feasible.

In addition, research and development in the computing sciences can profit from problem definitions and work procedures applied in forensics, e.g.,

- forensic data, skilled forgeries and partial, noisy data that pose *challenging problems* regarding the robustness of an automatic system.
- computer scientists can *gain new insights in analysis procedures* while taking the perspective of a forensic expert who has expertise in his / her field of specialization.
- *computational approaches undergo fine tuning* to achieve superiority, but eventually also generalization.

As a new scientific discipline, approaches and studies in CF need to be peer-reviewed and published for the purpose of discussion, consequent general acceptance, and rejection by the scientific community. Scientific expertise from forensics as well as computing have to be incorporated. Methods and studies have to be reviewed for their forensic and technological correctness. In addition, a legal framework needs to be established that deals with specific questions regarding the combined usage of human and machine intelligence in crime investigations. With the computer science background of the authors this framework can not be sketched comprehensibly. Yet, the following objections might inspire further discussions and studies.

- the digital representation of the trace evidence is insufficient and lacks particular detail information that can be observed in the original (analog) trace found at the crime scene (loss of information due to digitalization process).
- the extracted numerical parameters / features describe a particular detail of the trace insufficiently (loss of information due to inappropriate features).
- the applied computational method is not appropriate for a particular problem studied.

– the conclusions are misleading due to "wrong" results provided by the employed computational method, e.g., for classification, identification and verification.

Similar objections need to be answered in the classical forensic science already, in computational forensic, however, the perspective is broader; taking aspects of computer technology, methodology and application into account.

4   Computational / Machine Intelligence

Forensic methods can be assisted by algorithms and software from several areas in the computational science. Some of these are:

– *signal / image irocessing:* where one-dimensional signals and two-dimensional images are tranformed for the purpose of better human or machine processing,
– *computer vision:* where images are automatically recognized to identify objects,
– *computer graphics / data visualization:* where two-dimensional images or three-dimensional scenes are synthesized from multi-dimensional data for better human understanding,
– *statistical pattern recognition:* where abstract measurements are classified as belonging to one or more classes, e.g., whether a sample belongs to a known class and with what probability,
– *data mining:* where large volumes of data are processed to discover nuggets of information, e.g., presence of associations, number of clusters, outliers in a cluster,
– *robotics:* where human movements are replicated by a machine, and
– *machine learning:* where a mathematical model is learnt from examples.

Much of computational / machine intelligence is dominated by statistically based algorithms. These methods are ideally suited to forensics where there is a need to demonstrate error rates and calculate probabilities [7].

5   Application Examples

Mathematical, statistical and computer-based methods have been used before in forensics. Computer forensics (also called digital forensics) and DNA analysis are one example. Contributions to the *scientific methododological base* of handwriting and signature analysis are reported in [8,9], while *search algorithm* are proposed in [10]. For the *synthesis of data* samples not only software methods, but also robots are used [11]. Research on the computer-based analysis of *striation patterns* that are subject of ballistic / tool-mark investigations [12] is reported. *Friction ridge analysis* is probably the area that has most benefited from computational methods, with the development of automated fingerprint identification systems [13]. However much more needs to be done in the analysis of latent prints. Computer-assisted and fully automatic computer-based *link analysis and visualization* is increasingly used by banks and insurances in examining credit-card fraud and money laundry [3]. Crime-scene reconstruction using computer graphics referred to in [14]. A *conceptual framework on terminology used* by questioned document examiners is proposed [15] that was also implemented into a *reporting system* [16]. Assistance software for argumentation is discussed in [17]. The need for professionals with the abilities to develop and to apply latest computational methods demands *education and training* of current and next generation experts [14].

Computational forensic research generated a number of studies in the most recent years. Covered research topics and domains are diverse as for example information retrieval [18], data mining [19], digital forensics [20,21], device forensics [22], human identification (finger print [23] and speech recognition [24]), anthropology [25,26], linguistic [27,28], questioned documents [29], forensic statistics [30], and decision making [31].

6 Conclusions and Future Directions

The use of computing tools in the forensic disciplines is sometimes minimal. Many improvements in forensics can be expected if recent findings in applied mathematics, statistics and computer sciences are implemented in computer-based systems. The objectives of this paper were to:

i) increase awareness of the impact of computer tools in crime prevention, investigation and prosecution; on the one hand, among forensic scientists, e.g., with expertise in biology, chemistry and medicine but with limited exposure to computational science, and on the other hand, among computer scientists unaware of a challenging application domain.
ii) introduce computer scientists to the needs, procedures and techniques of forensics, and
iii) motivate studies on computational tools in forensics and encourage joint development by forensic and computer scientists.

With the introduction of computer-based methods in the investigation processes, the advancement of technology and methodology as depict in Figure 1, new work procedures and legal frameworks need to be established that take advantage of both knowledge domains; forensic and computational sciences.

Several support methods are needed for CF development: international forums (e.g. conference, scientific press media) to review and exchange research results, education and training to prepare current and future researchers and practitioners, and financial support for research and development.

Computational forensics holds the potential to greatly benefit all of the forensic sciences. For the computer scientist it poses a new frontier where new problems and challenges are to be faced. The potential benefits to society, meaningful inter-disciplinary research, and challenging problems should attract high quality students and researchers to the field.

References

1. van der Steen, M., Blom, M.: A roadmap for future forensic research. Technical report, Netherlands Forensic Institute (NFI), The Hague, The Netherlands (2007)
2. Saks, M., Koehler, J.: The coming paradigm shift in forensic identification science. Science 309, 892–895 (2005)
3. Mena, J.: Investigative Data Mining for Security and Criminal Detection. Butterworth-Heinemann (2003)
4. Starzecpyzel: United states vs. Starzecpyzel. 880 F. Supp. 1027 (S.D.N.Y) (1995)
5. Marr, D.: Vision. Freeman, New York (1982)
6. Foster, K., Huber, P.: Judging Science. MIT Press, Cambridge (1999)
7. Aitken, C., Taroni, F.: Statistics and the Evaluation of Evidence for Forensic Scientists, 2nd edn. Wiley, Chichester (2005)
8. Franke, K.: The Influence of Physical and Biomechanical Processes on the Ink Trace - Methodological foundations for the forensic analysis of signatures. PhD thesis, Art. Intell. Institute, Uni. Groningen, The Netherlands (2005)
9. Srihari, S., Cha, S., Arora, H., Lee, S.: Individuality of handwriting. J. Forensic Sciences 47(4), 1–17 (2002)
10. Bulacu, M., Schomaker, L.: Text-independent writer identification and verification using textural and allographic features. IEEE Trans. Pattern Analysis and Machine Intelligence (PAMI) 29(4), 701–717 (2007)
11. Franke, K., Schomaker, L.: Robotic writing trace synthesis and its application in the study of signature line quality. J. Forensic Doc. Examination 16(3) (2004)
12. Heizmann, M., León, F.: Model-based analysis of striation patterns in forensic science. In: Bramble, S., Carapezza, E., Rudin, L. (eds.) Enabling Technologies for Law Enforcement and Security, Proceedings of SPIE, vol. 4232, pp. 533–544 (2001)
13. Maltoni, D., Maio, D., Jain, A., Prabhakar, S.: Handbook of Fingerprint Recognition. Springer, Heidelberg (2003)
14. Veenman, C., Worring, M.: Forensic intelligence. Informatie, 60–65 (April 2007)
15. Franke, K., Guyon, I., Schomaker, L., Vuurpijl, L.: WandaML - A markup language for digital document annotation. In: Proc. 9th International Workshop on Frontiers in Handwriting Recognition (IWFHR), Tokyo, Japan (2004)
16. Schönherr, K.: Konzeption und Prototyp eines Ausgabe- und Reportgenerators für XML-Daten aus dem Handschriftenerkennungssystem WANDA. Master's thesis, Berlin College of Technology and Business Studies, Berlin, Germany (2004)
17. Verheij, B.: Virtual Arguments. On the Design of Argument Assistants for Lawyers and Other Arguers. T.M.C. Asser Press, The Hague, The Netherlands (2005)
18. Su, H.: Shoeprint image retrieval based on local image features. In: Int. Symposium on Information Assurance and Security/ Int. Workshop on Computational Forensics (IWCF), Manchester, UK. IEEE-CS Press, Los Alamitos (2007)
19. Bache, R., Crestani, F., Canter, D., Youngs, D.: Application of language models to suspect prioritisation and suspect likelihood in serial crimes. In: Int. Symposium on Information Assurance and Security/ IWCF. IEEE-CS Press, Los Alamitos (2007)
20. Veenman, C.: Statistical disk cluster classification for file carving. In: Int. Symposium on Information Assurance and Security/ IWCF. IEEE-CS Press, Los Alamitos (2007)
21. Karresand, M.: Completing the picture, fragments and back again (licentiate thesis) (May 2008)

22. Khanna, N., Mikkilineni, A., Chiu, G., Allebach, J., Delp, E.: Survey of scanner and printer forensics at purdue university. In: Srihari, S.N., Franke, K. (eds.) IWCF 2008. LNCS, vol. 5158, pp. 22–34. Springer, Heidelberg (2008)

23. Srihari, S., Srinivasan, H., Fang, G.: Discriminability of the fingerprints of twins. Journal of Forensic Identification 58(1), 109–127 (2008)

24. Weiand, K., Bouten, J., Veenman, C.: Similarity visualisation for the grouping of forensic speech recordings. In: Srihari, S.N., Franke, K. (eds.) IWCF 2008. LNCS, vol. 5158, pp. 169–180. Springer, Heidelberg (2008)

25. Ballerini, L., Cordon, O., Damas, S., Santamaria, J., Aleman, I., Botella, M.: Craniofacial superimposition in forensic identification using genetic algorithms. In: Int. Symposium on Information Assurance and Security/ IWCF. IEEE-CS Press, Los Alamitos (2007)

26. Ehlert, A., Bartz, D.: 3d processing and visualization of scanned forensic data. In: Srihari, S., Franke, K. (eds.) 2nd International Workshop on Computational Forensics (IWCF). LNCS, p. 70. Springer, Heidelberg (2008)

27. Hughes, D., Rayson, P., Walkerdine, J., Lee, K., Greenwood, P., Rashid, A., May-Chahal, C., Brennan, M.: Supporting law enforcement in digital communities through natural language analysis. In: Srihari, S.N., Franke, K. (eds.) IWCF 2008. LNCS, vol. 5158, pp. 122–134. Springer, Heidelberg (2008)

28. Booker, L.: Finding identity group "fingerprints" in documents. In: Srihari, S.N., Franke, K. (eds.) IWCF 2008. LNCS, vol. 5158, pp. 113–121. Springer, Heidelberg (2008)

29. van Beusekom, J., Shafait, F., Breuel, T.: Document signatures using intrinsic features for counterfeit detection. In: Srihari, S.N., Franke, K. (eds.) IWCF 2008. LNCS, vol. 5158, pp. 47–57. Springer, Heidelberg (2008)

30. Ramos, D., Gonzalez-Rodriguez, J., Zadora, G., Zieba-Palus, J., Aitken, C.: Information-theoretical comparison of likelihood-ratio methods of forensic evidence evaluation. In: Int. Symposium on Information Assurance and Security/ IWCF, Manchester, UK. IEEE-CS Press, Los Alamitos (2007)

31. Yanushkevich, S., Boulanov, O., Stoica, A., Shmerko, V.: Support of interviewing techniques in physical access control systems. In: Srihari, S.N., Franke, K. (eds.) IWCF 2008. LNCS, vol. 5158, pp. 147–158. Springer, Heidelberg (2008)

Computational Methods for Determining Individuality

Sargur N. Srihari and Chang Su

Center of Excellence for Document Analysis and Recognition (CEDAR)
University at Buffalo, State University of New York
Amherst, New York 14228, U.S.A
{srihari,changsu}@cedar.buffalo.edu

Abstract. Individuality is the state or quality of being an individual. We establish a computational methodology to determine whether a particular modality of data is sufficient to establish the individuality of every individual or even a demographic group. To test the individuality, generative models are given or learned to represent the distribution of certain characteristics such as birthday, human heights and fingerprints. Given the individuality assessments of different characteristic, the models based on multiple characteristics are proven to get strengthen individuality.

Keywords: individuality, generative model, fingerprint.

1 Introduction

As commonly used, individual refers to a person or to any specific object in a collection. From the seventeenth century on, individual indicates separateness, as in individualism. Individuality is the state or quality of being an individual; a person separate from other persons. Individuality study is important in forensic identification since we need to assess whether a particular input such as gender, height, weight or fingerprints can be used to identify specific person from the trace evidence they leave, often at a crime scene or the scene of an accident.

Studies on the individuality started from the late 1800s. About 20 models have been proposed since then trying to establish the improbability of two random people having the same certain characteristics. All the models try to quantify the uniqueness property to be able to defend forensic identification as a legitimate proof of identification in the courts. Each of these models try to find out the probability of false correspondence, i.e. probability that a wrong person is identified given a certain evidence collected from a crime scene from a set of previously recorded whole database. i.e., the probability that the features of two fingerprints match though they are taken from different individuals. A match here does not necessarily mean an exact match but a match within given tolerance levels. All the models establish the probability of two different people being identified as the same based on their features, namely, the probability of random correspondence (PRC). The models have been classified based on the different approaches that have been taken through a century of individuality studies. The

S.N. Srihari and K. Franke (Eds.): IWCF 2008, LNCS 5158, pp. 11–21, 2008.

latest class of models is called generative models. Generative models are statistical models that represent the distribution of the feature. In these models, a distribution of the features is learnt through a training dataset. Features are then generated from this distribution to test their individuality. What training set is used is immaterial as long as it is representative of the entire population.

The following of this paper is organized as follows: We discuss individuality computational method of birthday in Section 2 and heights in Section 3. Section 4.1 introduces a new generative model for both minutiae and ridges and individuality computation are also given. The paper concludes with a summary in Section 5.

2   Individuality of Birthday

Generative models for determining individuality can be understood by considering the trivial example of using the birthday of a person. The birthday problem asks whether any of the k people have a matching birthday with any of the others.

To compute the approximate probability that in a room of k people, at least two have the same birthday, we disregard variations in the distribution, such as leap years, twins, seasonal or weekday variations, and assume that the 365 possible birthdays are equally likely. Therefore the PRC value for birthday problem is 1/356. Uniform density is used here to model the birthday distribution.

It is easier to first calculate the probability $p(k)$ that all n birthdays are different. If $k > 365$, by the pigeonhole principle this probability is 0. On the other hand, if $k \leq 365$, it is by the pigeonhole principle this probability is 0. On the other hand, if k 365, it is

$$\bar{p}(k) = 1 \times \left(1 - \frac{1}{365}\right) \times \left(1 - \frac{2}{365}\right) \cdots \left(1 - \frac{k-1}{365}\right) = \frac{365!}{365^k(365-k)!} \quad (1)$$

The event of at least two of the k persons having the same birthday is complementary to all k birthdays being different. Therefore, its probability $p(k)$ is

$$p(k) = 1 - \bar{p}(k) \quad (2)$$

This probability surpasses 1/2 for $k = 23$ (with value about 50.7%).

3   Individuality of Human Height

The goal of the generative model for height, is to come up with an analytical value for the probability of two individuals having the same height within some tolerance $\pm \epsilon$. Different with the birthday problem, PRC value can not be computed directly. We have to learn the parameters of the generative model firstly. The steps in studying individuality using a generative model are given below.

1. Consider a probabilistic generative model and estimate its parameters from a particular data set.

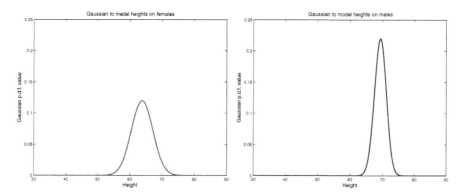

Fig. 1. Gaussian density used to model heights of individuals $\mu = 5.5$ and $\sigma = 0.5$ i.e. mean 5.5 feet and standard deviation 6 inches

2. Evaluate analytically the probability of two individuals to have the same height(or other bio-metric), with some tolerance $\pm\epsilon$.

For the study of individuality of height, a Gaussian density is a reasonable model to fit the distribution of heights of individuals. The height statistics is collected from CDC Advance Data No. 361 [?]. Figure 1 shows modeling the heights (inch) for males and females aged 20 years and over using a Gaussian p.d.f. with mean $\mu_f = 63.8$, $\mu_m = 69.3$ and standard deviation $\sigma_f = 11.1$, $\sigma_m = 3.3$. Now the probability of two individuals having the same height with some tolerance $\pm\epsilon$ can be derived as follows.

Probability of one individual having height $a \pm \epsilon$ is $\displaystyle\int_{a-\epsilon}^{a+\epsilon} P(h|\mu,\sigma)dh$

where $\quad P(h|\mu,\sigma) \sim \mathcal{N}(\mu,\sigma) = \dfrac{1}{\sqrt{2\pi}\sigma}e^{-\frac{(h-\mu)^2}{2\sigma^2}}.$

Probability of two individuals having height $a \pm \epsilon$ is $\quad \left(\displaystyle\int_{a-\epsilon}^{a+\epsilon} P(h|\mu,\sigma)\right)^2$

Probability of two individuals having *any* same height $\pm\epsilon$ is

$$p_\epsilon = \int_{-\infty}^{\infty}\left(\int_{a-\epsilon}^{a+\epsilon} P(h|\mu,\sigma)dh\right)^2 da \qquad (3)$$

Eq 3 can be numerically evaluated for a given value of μ,σ. Figure 2 shows the probability values for fixed $\mu_f = 63.8$, $\mu_m = 69.3$ and varying σ_f, σ_m. A tolerance of 0.1 inches was used in the probability calculations. It is obvious to note that, when σ decreases, the width of the Gaussian is smaller and hence the probability that two individuals having the same height is more.

The Probability of Random Correspondence for female height with a mean height of 63.8 inch and standard deviation of 11.1 inches is $p_\epsilon = 0.0025$. i.e. 25

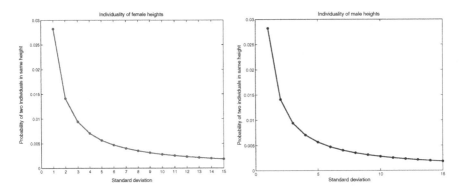

Fig. 2. Individuality of heights calculated using a Gaussian as a generative model. For different values of σ and fixed $\mu_f = 63.8$, $\mu_m = 69.3$, the probabilities are calculated.

out of every 10000 female have the same height (tolerance of 0.1 inches) and the Probability of Random Correspondence for male height with a mean height of 69.3 inch and standard deviation of 3.3 inches is $p_\epsilon = 0.0085$. i.e. 85 out of every 10000 men have the same height (tolerance of 0.1 inches)

Based on this model, the probability of at least two people sharing a height among k individuals can be estimated. This evaluation is similar to that of the birthday problem where the probability of two people having the same birthday among k individuals is calculated. In the case of heights we have a real value instead of 365 date value. This is handled by the implicit discreteness due to the tolerance ϵ.

Assuming that there are h possible height and all h possible heights are equally likely, $\bar{p}(k)$ is given by

$$\bar{p}(k) = 1 \cdot \left(1 - \frac{1}{h}\right) \cdot \left(1 - \frac{2}{h}\right) \cdots \left(1 - \frac{k-1}{h}\right) = \frac{h!}{h^k(h-k)!} \qquad (4)$$

The event of at least two of the k persons having the same height is complementary to all k heights being different. Therefore, its probability $p(k)$ is

$$p(k) = 1 - \bar{p}(k) \qquad (5)$$

Thus $p(2) = 1 - \bar{p}(2) = 1 - 1 \cdot (1 - \frac{1}{h})$, since $p_\epsilon = p(2)$ we have $h = 1/p_\epsilon$.

The Table 1 shows the probability of at least two people sharing a same birthday or height amongst a certain number of people, assuming $p_\epsilon = 0.025$. In a group of 10 (or more) randomly chosen people, there is more than 11% probability that some pair of them will have the same birthday and 10% for female and 32% for male probability that some pair of them will have the same height. For 80 or more people, the probability is more than 99%, tending toward 100% as the pool of people grows.

Table 1. The probability of at least two people sharing a same birthday or height amongst a certain number of people

k	p(k) Birthday	Female Heights	Male Heights
2	0.0028	0.0025	0.0085
5	0.0277	0.0248	0.0825
10	0.1194	0.1072	0.3251
20	0.4184	0.3830	0.8196
40	0.8966	0.8670	0.9995
80	0.9999	0.9998	$1 - 4 \times 10^{-13}$
120	$1 - 1.9 \times 10^{-6}$	$1 - 2.2 \times 10^{-6}$	1
200	$1 - 2.7 \times 10^{-11}$	$1 - 3.1 \times 10^{-11}$	1
300	$1 - 7.0 \times 10^{-22}$	$1 - 7.2 \times 10^{-22}$	1
400	1	1	1

4   Individuality of Fingerprints

Fingerprints have been used for identification from the early 1900s. Their use for uniquely identifying a person has been based on two premises, that, (i) they do not change with time and (ii) they are unique for each individual. Until recently, fingerprints had been accepted by courts as a legitimate proof of identification. But, after the 1999 case *US vs Byron Mitchell*, fingerprint identification has been challenged under the basis that the premises stated above have not been objectively tested and the error rates have not been scientifically established. Though the first premise has been accepted, the second one on individuality is the widely challenged one.

4.1   Generative Models for Fingerprint Individuality

Studies on the individuality of fingerprints date back to the late 1800s. All previous models can be classified into five different categories, namely, grid-based models, ridge-based models, fixed probability models, relative measurement models and generative models. Grid-based models include Galton [1] and Osterburgh [2] which were proposed in the late 80s and the early 90s respectively. One instance of ridge-based models is introduced by Roxburgh [3]. Fixed probability models contain the class of Henry-Balthazard models [4]. Relative measurement models include the Champod model [5] and the Trauring model [6]. The latest class of models, namely, the generative models aim at being flexible to represent observed distributions through different fingerprint databases and then ascertained uncertainties from models. Based on the the assumed non-independence of minutia locations and orientations, various mixture models could be used [7] and [8].

   In existing generative models only minutiae have been modeled without considering ridge features. Minutiae means small details in the fingerprints, it refers to the ridge endings and ridge bifurcation. See Figure 3 for the examples of

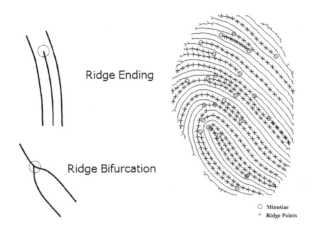

Fig. 3. (a) Minutiae: ridge ending and ridge bifurcation (b) detected minutiae and ridge points on a skeleton fingerprint image

two kinds of minutiae and the minutiae on a skeleton fingerprint image. We further embed ridge information into existing generative models by using the distribution for ridge points. The proposed model offers more reasonable and accurate fingerprint representation and therefore a more reliable probability of random correspondence (PRC). In this model, the ridge is represented as a set of ridge points sampled at equal interval of inter ridge width. Ridge length is defined as the number of ridge points that could be sampled from the ridge. Three types of ridges are defined as (i) short ridges: $l(r) \leq L/3$, (ii) medium ridges: $L/3 < l(r) < 2L/3$ and (iii) long ridges: $2L/3 \leq l(r) \leq L$, where L is the maxima ridge length which was collected from the FVC 2002 database.

These three possible ridge length types can be associated with any minutiae. Without loss of generality, we can assume that there exist only three possible ridge length types corresponding to a minutiae. For the generative model, the ridge length type is modeled as a uniform distribution $F^l(l_r|a, b)$, where $[a, b]$ is the interval of the uniform distribution. For ridges with different lengths, different ridge points are picked as anchors. The index to be used for ridge point selection should satisfy following two conditions: 1) The index should be large so as to infer as many other ridge points as possible. 2) The index should not be too large to overstep the ridge length. A tradeoff has to be balanced between the two conditions [9]. For medium ridges, $(L/3)^{th}$ ridge point is picked and for long ridges, both $(L/3)^{th}$ and $(2L/3)^{th}$ are picked. None ridge point will be chosen for short ridges. For the generative model, the ridge points are modeled as a combining distribution of the ridge point location and the direction. The proposed joint distribution model for fingerprint presentation is based on a mixture consisting of G components. Each components is distributed according the density of the minutiae and the ith ridge points: $F_g^m \, F_g^i$. The equation of the generative model is given in Eq. 6.

$$f(\cdot|\Theta_G) = \begin{cases} F^l(l_r) \cdot \sum_{g=1}^{G_1} \pi_g F_g^m(s_m, \theta_m|\Theta_G) & l_r \leq L/3 \\[2ex] F^l(l_r) \cdot \sum_{g=1}^{G_2} \pi_g F_g^m(s_m, \theta_m|\Theta_G) \\ \cdot F_g^{L/3}(r_{L/3}, \phi_{L/3}, \theta_{L/3}|\Theta_G) & L/3 < l_r < 2L/3 \\[2ex] F^l(l_r) \cdot \sum_{g=1}^{G_3} \pi_g F_g^m(s_m, \theta_m|\Theta_G) \cdot F_g^{L/3}(r_{L/3}, \phi_{L/3}, \theta_{L/3}|\Theta_G) \\ \cdot F_g^{2L/3}(r_{2L/3}, \phi_{2L/3}, \theta_{2L/3}|\Theta_G) & l_r \geq 2L/3 \end{cases} \tag{6}$$

In Eq. 6, $F_g^m(\cdot)$ represents the distribution of the minutiae location s_m and the direction θ_m. $F_g^i(\cdot)$ presents the distribution of the ith ridge points. To estimate the unknown parameters in the generative model, we develop an algorithm based on the EM algorithm. Different numbers of the components G for the mixture model were validated using k-means clustering. The one with the best k-means clustering results was chosen.

4.2   Fingerprint Individuality Computation

Given a template T with n minutiae and an input/query Q with m minutiae and corresponding ridge points pairs and w out of them match, the probability of Random Correspondence is given by

$$PRC_0 = p^*(w; Q, T) =$$
$$= \binom{n}{w} \cdot (p_m(Q, T))^w (1 - p_m(Q, T))^{n-w} \tag{7}$$

The probability is a binomial probability whose parameters are n and $p_m(Q, T)$. The latter is the probability that a random minutiae and corresponding ridge points pair from Q will match a pair from T. Since most of the matchers try to maximize the number of matchings (i.e. they would find a matching even in a fingerprint that are totally different, we calculate the conditional expectation, conditioned on that fact that the number of matches is always greater than zero and equating this to the number of pair matches between Q and T, the estimation can be written as

$$\frac{n.p_m(Q, T)}{(1 - (1 - p_m(Q, T))^n)} = w_0 \tag{8}$$

w_0 is found out by the proposed models fit into Q and T and determining the number of matches by $k-plet$ [10] matching algorithm. Value of $p_m(Q, T)$ can be found from Eq 8. In a database contains N different fingers with L impressions of the same finger, $(Q, T)_{impostor}$ is used to denote all the $N(N-1)L^2/2$ impostor pairs.

$$\overline{PRC} = \frac{1}{N(N-1)L^2/2} \sum_{(Q,T)_{impostor}} p^*(w; Q, T) \tag{9}$$

4.3   Experiments and Results

Generative models without ridge information and with the ridge information model introduced in 4.1 have been implemented and experiments have been conducted on FVC2002 DB1 [11]. The number of components G for the mixture model was found after validation using k-means clustering. The database has 100 different fingerprints with 8 impressions of the same finger. Thus, there are a total of 800 fingerprints using which the model has been developed.

Table 2. PRC for different fingerprint matches with varying m (number of minutiae in template),n (number of minutiae in input) and w (number of matched minutiae or minutiae and corresponding ridge points pairs) - With ridge information and without ridge information. **PRC_0** is PRC for the general population and $\overline{PRC}$ is for PRC for FVC2002-DB1.

m	n	w	With Ridge Information and Minutiae		With Only Minutiae	
			PRC_0	$\overline{PRC}$	PRC_0	$\overline{PRC}$
16	16	4	3.9×10^{-2}	1.6×10^{-3}	2.1×10^{-1}	2.1×10^{-1}
		8	1.8×10^{-5}	1.7×10^{-8}	1.1×10^{-2}	7.8×10^{-3}
		16	8.9×10^{-18}	3.1×10^{-24}	4.8×10^{-11}	1.6×10^{-11}
26	26	6	7.4×10^{-3}	7.9×10^{-4}	1.3×10^{-1}	1.4×10^{-1}
		12	6.9×10^{-8}	3.8×10^{-10}	3.6×10^{-4}	5.4×10^{-4}
		20	2.3×10^{-18}	2.4×10^{-22}	2.3×10^{-11}	5.3×10^{-11}
		26	2.2×10^{-30}	1.2×10^{-35}	6.7×10^{-21}	2.1×10^{-20}
36	36	6	1.8×10^{-2}	4.1×10^{-3}	1.5×10^{-1}	1.7×10^{-1}
		16	1.4×10^{-10}	8.5×10^{-13}	5.1×10^{-6}	2.8×10^{-5}
		26	1.1×10^{-23}	1.6×10^{-27}	1.6×10^{-15}	4.2×10^{-14}
		36	8.7×10^{-44}	3.6×10^{-49}	5.6×10^{-32}	7.3×10^{-30}
46	46	6	2.6×10^{-2}	1.0×10^{-2}	1.6×10^{-1}	1.6×10^{-1}
		20	7.8×10^{-14}	7.4×10^{-16}	5.2×10^{-8}	9.8×10^{-7}
		32	4.8×10^{-30}	2.0×10^{-33}	6.6×10^{-20}	1.5×10^{-17}
		46	9.9×10^{-59}	1.0×10^{-63}	1.4×10^{-43}	6.1×10^{-40}

We compare the results to that of [8]. Random fingerprints are generated from the model. Values of PRC_0 and $\overline{PRC}$ are calculated using the formulae introduced in Section 4.1. The results are presented in Table 2. The PRCs are calculated through varying number of minutiae in template(m), Input(n) and the number of ridges matched(w). Table 2 shows that more the number of minutiae in the template and the input, the higher the PRC. In experiments conducted on the FVC2002 DB1, there are some differences between the results obtained here and the results in Jain et al. [8]. This may result from use of different matching algorithms, which w_0 depends on. Our highlight is that the PRC values embedded with ridge information model are never greater than PRC values without ridge information. Table 2 also shows the PRC values corresponding to use of ridge information model in the generative model and these probabilities are lesser when compared to those without ridge information which indicates that ridge information strengthens individuality of fingerprints. The PRCs for

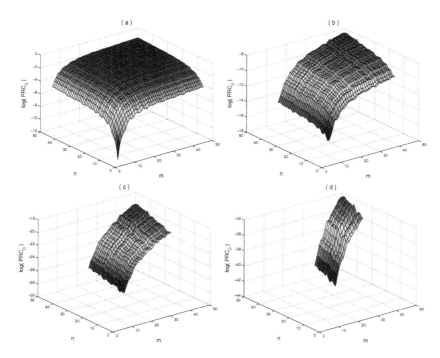

Fig. 4. PRCs with different number of the matched ridges for (a) $w = 6$, (b) $w = 16$, (c) $w = 26$, and (d) $w = 36$

Table 3. The probability of at least two fingerprints matched among a certain number of fingerprints with average number of minutiae $m = n = 39$ and average number of matched minutiae $w = 27$

k	Minutiae and Ridge Information p(k)	only Minutiae p(k)
2	1.54×10^{-24}	1.18×10^{-15}
5	1.54×10^{-23}	1.18×10^{-14}
10	6.93×10^{-23}	5.31×10^{-14}
100	7.64×10^{-21}	5.84×10^{-12}
10^5	7.72×10^{-15}	5.90×10^{-6}
3.03×10^8	7.09×10^{-8}	1.12×10^{-2}
6.66×10^9	3.42×10^{-5}	6.77×10^{-2}
10^{10}	7.72×10^{-5}	7.02×10^{-1}
8.48×10^{14}	1.03×10^{-1}	1
10^{20}	0.999999999999999946	1
6.48×10^{23}	1	1

the different m and n with 6, 16, 26 and 36 matching ridges are shown in Figure 4. It is obvious to note that, when w decreases or m and n increase, the probability that two random fingerprints matching is more.

Because the ridge information models are independent on generative models, other recently proposed generative models such as [12] could also be embedded with ridge information in similar manner and are also expected to offer more reliable PRC values.

The probability of at least two fingerprints matched among a certain number of fingerprints is computed similarly as Eq. 5 as well and given by Table 3. In 100,000 randomly chosen fingerprints, there is only 7.72×10^{-15} probability that some pair of them will match if we consider both minutiae and ridge in matching. This probability is much smaller than previous minutiae only model which is 5.90×10^{-6}. The probability of at least two fingerprints matched among U.S. and world population are 7.09×10^{-8} and 3.42×10^{-5} respectively.

5 Summary

Generative models of individuality attempt to model the distribution of features and then use the models to determine the probability of random correspondence. This paper provides a detailed survey of individuality models. We have analyzed 3 models based on birthday, heights and fingerprints. For birthday model, generative model fits uniform distribution and PRCs can be gotten directly. Human height model uses Gaussian density to present the height distribution. PRCs are computed from the generative model leaned form statistic data. Fingerprint individuality computation is the most complicated one. A generative model with an mixture distribution is used to model both minutiae and ridge information. The new generative models are learned and then compared by the experiments with the generative model without ridge information on the FVC2002 DB1. The PRCs obtained for a fingerprint template and input with 36 minutiae each with 16 matching minutiae is 1.4×10^{-10} (or 14 in a 100,000 million, or equivalently, 1 in 7,000 million). This is a much stronger result than without using ridge information which is 1 in 200,000. With 20 matching minutiae this probability is one in 300 trillion, as opposed to the earlier result of 1 in 100 million in [8]. Since proposed ridge information model offers a more reasonable and more accurate fingerprint representation, PRC values with ridge information are much smaller than PRC values without ridge information.

Acknowledgments. This work was supported by a grant from the Department of Justice, Office of Justice Programs Grant NIJ 2005-DD-BX-K012. The opinions expressed are those of the authors and not of the DOJ.

References

1. Galton, F.: FingerPrints. McMillan, London (1892)
2. Osterburg, J.: Development of a mathematical formula for the calculation of fingerprint probabilities based on individual charectiristics. Journal of American Statistical Association 772, 72 (1997)
3. Roxburgh, T.: Galton's work on the evidential value of fingerprints. Indian Journal of Statistics 1, 62 (1933)

4. Henry, E.: Classification and Uses of FingerPrints. Routledge & Sons, London (1900)
5. Champod, C., Margot, P.: Computer assisted analysis of minutiae occurrences on fingerprints. In: Almog, J., Spinger, E. (eds.) Proc. International Symposium on Fingerprint Detection and Identification, Israel National Police, Jerusalem, p. 305 (1996)
6. Trauring, M.: Automatic comparison of finger-ridge patterns. Nature, 197 (1963)
7. Pankanti, S., Prabhakar, S., Jain, A.K.: On the individuality of fingerprints. IEEE Transactions on Pattern Analysis and Machine Intelligence 24(8) (2002)
8. Dass, S., Zhu, Y., Jain, A.K.: Statistical models for assessing the individuality of fingerprints. In: Fourth IEEE Workshop on Automatic Identification Advanced Technologies, pp. 3–9 (2005)
9. Fang, G., Srihari, S.N., Srinivasan, H., Phatak, P.: Use of ridge points in partial fingerprint matching. In: Proc. of SPIE: Biometric Technology for Human Identification, vol. IV, pp. 65390D1–65390D9 (2007)
10. Chikkerur, S., Cartwright, A.N., Govindaraju, V.: K-plet and cbfs: A graph based fingerprint representation and matching algorithm. In: International Conference on Biometrics, pp. 309–315 (2006)
11. Maio, D., Maltoni, D., Cappelli, R.: Fingerprint verification competition (2002), http://bias.csr.unibo.it/fvc2002/
12. Zhu, Y., Dass, S., Jain, A.: Compound stochastic models for fingerprint individuality. In: Proc. of International Conference on Pattern Recognition, vol. 3, pp. 532–535 (2006)

Survey of Scanner and Printer Forensics
at Purdue University

Nitin Khanna, Aravind K. Mikkilineni,
George T.-C. Chiu, Jan P. Allebach, and Edward J. Delp*

Purdue University, West Lafayette IN 47907, USA**
*ace@ecn.purdue.edu

Abstract. This paper describes methods for forensic characterization of scanners and printers. This is important in verifying the trust and authenticity of data and the device that created it. An overview of current forensic methods, along with current improvements of these methods is presented. Near-perfect identification of source scanner and printer is shown to be possible using these techniques.

1   Introduction

Advances in digital imaging technologies have led to the development of low-cost and high-resolution digital cameras and scanners, both of which are becoming ubiquitous. Digital images generated by various sources are widely used in a number of applications from medical imaging and law enforcement to banking and daily consumer use. The increasing functionality of image editing software allows even an amateur to easily manipulate images. In some cases a digitally scanned image can meet the threshold definition requirements of a "legal duplicate" if the document can be properly authenticated. Forensic tools that help establish the origin, authenticity, and the chain of custody of such digital images are essential to a forensic examiner.

The same holds true for printed material, which in many cases is a direct accessory to many criminal and terrorist acts. Examples include forgery or alteration of documents used for purposes of identity, security, or recording transactions. In addition, printed material may be used in the course of conducting illicit or terrorist activities. Examples include instruction manuals, team rosters, meeting notes, and correspondence. In both cases, the ability to identify the specific device or type of device used to print the material in question would provide a valuable aid for law enforcement and intelligence agencies.

There are various levels at which this source identification problem can be addressed. One may want to find the particular device which generated the

---

* Corresponding author.
** This material is based upon work supported by the National Science Foundation under Grant No. CNS-0524540. Any opinions, findings, and conclusions or recommendations expressed in this material are those of the author(s) and do not necessarily reflect the views of the National Science Foundation.

image or one might be interested in knowing only the make and model of the device.

As summarized in [1], a number of robust methods, involving characterization of sensor noise, have been proposed for source camera identification. In [2, ?], techniques for classification of images based on their sources, scanned, camera generated, and computer generated, are presented.

Forensic characterization of a printer involves finding intrinsic features in the printed document that are characteristic of that particular printer, model, or manufacturer's products. This is referred to as the *intrinsic signature*. The intrinsic signature requires an understanding and modeling of the printer mechanism, and the development of analysis tools for the detection of the signature in a printed page with arbitrary content [4].

In this paper we present a brief overview of some recent advances in source identification of both scanned images and printed documents. Results are presented for our methods based on intrinsic signatures of the devices.

2   Characterization of Scanners

Pioneering work in the development of source camera identification was performed by Lukas et al. using techniques involving the imaging sensor's pattern noise [5,6]. Our method extends the use of sensor noise for source scanner identification by replacing correlation detector by statistical features and support vector machine (SVM) classifier. Since sensor pattern noise is estimated using the simple averaging method, further improvements in results may be obtained by using the improved method for sensor noise estimation presented in [6]. Extensive experimentation on a large set of scanners and different scanning scenarios show the effectiveness of our proposed scheme.

2.1   Statistical Feature Extraction

The basic structure of the scanner imaging pipeline and various types of noises involved in the scanning process are explained in [7]. Pattern noise refers to any spatial pattern that does not change significantly from image to image. It has been successfully used for source camera identification in [5,6]. In this study the sensor noise is modeled as the sum of two components, a random component and a fixed component. The random component is that portion of the noise which changes from image to image and varies over a period of time, while the fixed component is that portion of the noise which remains constant from image to image. The fixed component can be considered as a signature of the imaging sensor and can be used for source scanner identification. The challenge is to separate the fixed component from the random component of the noise, and then design suitable classifiers based on appropriate features. Since both CCD based and CIS based scanners suffer from similar noise sources, the proposed scheme is expected to work on both types of scanners. This is demonstrated by extensive experimentation.

Selection of relevant features from the sensor noise is the key to accurate and robust source scanner identification. The features selected should satisfy the following requirements:

- Independent of image content
- Capture the characteristics of a particular scanner and if possible differ amongst scanners of the same make and model
- Independent of scan area, that is, they should be able to characterize the source scanner even if the images are placed at different positions on the scanner's glass plate

An image scanned twice with the same scanner, but at different non-overlapping locations on the scanner bed will contain different PRNU due to variations in the manufacturing process such as doping concentration. The proposed scheme uses PRNU based scanner fingerprints. The fixed component of sensor noise is caused by PRNU as well as noise-like characteristics remaining after the post processing steps. To generate the final image, a number of non-linear operations are performed on the values read by the sensor array. Thus, the statistical properties of the fixed component of noise are expected to remain same irrespective of the image place-ment on the scanner bed. This is the reason behind using the statistical features of the fixed component of sensor noise for source scanner identification. Our experi-mental results suggest that this is true.

The scanned images are generated by translating a linear sensor array along the length of a document or image. Each row of the resulting digital image is generated by the same set of sensor pixels. Thus, for scanned images, the average of all the rows of sensor noise will give an estimate of the fixed "row-pattern". Averaging will reduce the random component while at the same time enhancing the fixed component of the noise. Along with using different statistical features of the "row-pattern", similar features are estimated along the column direction as well for comparing them with statistics along the row direction.

The procedure to extract features from one of the color channels is described here and is applied to all three channels separately to get the complete feature vector. Let I denote the input image of size $M \times N$ pixels (M rows and N columns) and I_{noise} be the noise corresponding to the image. Let $I_{denoised}$ be the result of applying a denoising filter on I. Then as in [5], $I_{noise} = I - I_{denoised}$. Let $\widetilde{I}^r_{noise}$ and $\widetilde{I}^c_{noise}$ denote the average of all the rows and the columns of the noise (I_{noise}), respectively. Further, let $\rho_{row}(i)$ denote the value of correlation between the average of all the rows ($\widetilde{I}^r_{noise}$) and the ith row of the noise (I_{noise}). Similarly, $\rho_{col}(j)$ denotes the value of correlation between the average of all the columns ($\widetilde{I}^c_{noise}$) and j^{th} column of the noise (I_{noise}) (Equation 1 for rows and similar for columns).

$$\widetilde{I}^r_{noise}(1,j) = \frac{1}{M}\sum_{i=1}^{M} I_{noise}(i,j); \qquad \rho_{row}(i) = \mathbf{C}(\widetilde{I}^r_{noise}, I_{noise}(i,.)) \qquad (1)$$

ρ_{row} is expected to have higher values than ρ_{col} since there is a periodicity between rows of the fixed component of the sensor noise of a scanned image.

The statistical properties of ρ_{row}, ρ_{col}, $\widetilde{I}^r_{noise}$ and $\widetilde{I}^c_{noise}$ capture the essential properties of the image which are useful for discriminating between different scanners. The mean, standard deviation, skewness and kurtosis of ρ_{row} and ρ_{col} are the first eight features, extracted from each color channel of the input image. The standard deviation, skewness and kurtosis of $\widetilde{I}^r_{noise}$ and $\widetilde{I}^c_{noise}$ correspond to features 9 through 14. The last feature for every channel of the input image is given by (2) which is representative of relative difference in periodicity along row and column directions in the sensor noise. Values of ρ_{col} corresponds to the case when each element is generated independently and so f_{15} will have high positive value. Some images from low-quality scanners or images which have undergone post-processing operations such as high down-sampling or JPEG compression, may be an exception to this.

$$\mathbf{f}_{15} = \left(1 - \frac{\frac{1}{N}\sum_{j=1}^{N}\rho_{col}(j)}{\frac{1}{M}\sum_{i=1}^{M}\rho_{row}(i)}\right) * 100 \qquad (2)$$

By extracting these 15 features from each of the three color channels, a 45 dimensional feature vector is obtained for each scanned image. To capture the three color channels, some scanners use three different linear sensors while others use a single imaging sensor in coordination with a tri-color light source. To capture this difference among scanners of different make and models, six additional features are used. These features are obtained by taking mutual correlations of $\widetilde{I}^r_{noise}$ for different color channels (same for $\widetilde{I}^c_{noise}$). Hence, in total each scanned image has a 51 dimensional feature vector associated with it.

In our previous work [7] on source scanner identification from images scanned at native resolution of the scanners, a recently developed anisotropic local polynomial estimator for image restoration based on directional multiscale optimizations [8] was used for denoising. In this study, a denoising filter bank comprising of four different denoising algorithms: LPA-ICI denoising scheme [8], median filtering (size 3×3) and Wiener adaptive image denoising for neighborhood sizes 3×3 and 5×5, is used. Using a set of denoising algorithms helps to better capture different types of sensor noises. These denoising algorithms are chosen based on the performance of the complete filter bank in scanner identification. Each denoising algorithm is independently applied to each color band of an image. The features extracted from individual blocks of the filter bank are concatenated together to give the final feature vector for each scanned image. Hence, each scanned image has a 204 dimensional feature vector associated with it. To reduce the dimensionality of the feature vectors, linear discriminant analysis (LDA) is used and a ten dimensional feature vector is obtained for each image. Each component of the ten dimensional feature vector is then a linear combination of the original 204 features. Finally a SVM classifier is trained and tested on these ten dimensional feature vectors.

2.2   Experimental Results

The following set of scanners are used in our experiments (label, make/model, DPI): { {S_1, Epson Perfection 4490 Photo, 4800}, {S_2, HP 6300c-1, 1200}, {S_3,

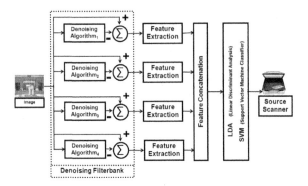

Fig. 1. Feature-based scanner identification

HP 6300c-2,1200}, $\{S_4$, HP 8250, 4800\}, $\{S_5$, Mustek 1200 III EP, 1200\}, $\{S_6$, Visioneer OneTouch 7300, 1200\}, $\{S_7$, Canon LiDe 25, 1200\}, $\{S_8$, Canon Lide 70, 1200\}, $\{S_9$, OpticSlim 2420, 1200\}, $\{S_{10}$, Visioneer OneTouch 7100, 1200\}, $\{S_{11}$, Mustek ScanExpress A3, 600\}\}. Experiments are performed on images scanned at the native resolution of the scanners as well as on images scanned at lower resolution, such as 200 DPI. This paper presents only results for 200 DPI images since images are generally scanned at lower resolution to meet constraints on storage space, scanning time, and transmission bandwidth. Experiments on native resolution images show similar accuracies.

The experimental procedure for source scanner identification using statistical features of sensor noise is shown in Fig. 1. The features are mapped to the range $[-1, 1]$ before SVM classification to achieve higher classification accuracy. The mapping is decided by the values of features in the training set and same mapping is applied on the features in the testing set. A radial basis function is chosen as the kernel function and a grid search is performed to select the best parameters for the kernel. To generate the final confusion matrices, SVM training and testing steps are repeated multiple times using a random selection of training and testing sets.

In the next few experiments, the effectiveness of the proposed scheme is shown for heavily sub-sampled (200 DPI) images. The scheme proposed here has good performance on 200 DPI images (which corresponds to scaling by 17% to 4% for native resolutions of 1200 DPI to 4800 DPI). 108 images are scanned at 200 DPI using each of the eleven scanners. All these scanned images are saved as uncompressed TIFF images. Unless stated otherwise, for each experiment 80 randomly selected images from each class are used for training and the remaining images are used for testing.

Degradation in the characteristics of sensor noise due to heavy down-sampling prevents successful separation of images scanned from the two scanners of exact same make and model as demonstrated by our initial experiments. For all further experiments performed on images scanned at 200 DPI, scanners S_2 and S_3 are

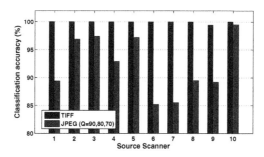

Fig. 2. Classification accuracies of multi-class classifiers. (Dedicated classifier for TIFF and general classifier for JPEG).

treated as a single class. The bar graph shown in Fig. 2 shows the classification accuracies of the multi-class classifier for 200 DPI TIFF images. The proposed algorithm has an average classification accuracy of 99.9% among ten scanner models.

To further check the robustness of the proposed scheme for scanner model identification, a SVM classifier is trained without images from scanner S_3 and tested on images from the scanner S_3 only. This has classification accuracy of 95% for the testing images from scanner S_3. A similar classifier designed by training without images from scanner S_2 and testing on images from S_2 only, gives a classification accuracy of 97%. These results imply that even in the absence of the training data from a particular scanner, the proposed scheme can identify the scanner model as long as training data for another scanner of the same make and model is available.

Another aspect of robustness is independence from scanning location. That is, even when the image is placed at a random unknown location on the scanner bed, source scanner identification should still be possible. The images used in all the earlier experiments were scanned from the "default" scanning location (generally marked at the top left corner) of the scanner. For this experiment another 108 images are scanned from scanner S_{11}, with their location on the scanner's bed slightly translated horizontally and vertically between each scan. A SVM classifier is trained using the images scanned from "default" location and tested using images from the "random" locations only. This has a classification accuracy of 100% for "randomly" placed images. This suggests that the proposed scheme for scanner model identification is independent of the scanning location.

To investigate the robustness of the proposed scheme under JPEG compression, TIFF images from all the scanners are compressed at three different quality factors $Q = 90$, 80 and 70. For using a dedicated classifier on distorted images, we need to know the particular post-processing applied on the image of unknown origin. In some cases, this a priori information is available or can be obtained by using other forensic methods. For example, it may be possible to know the JPEG quality factor by an analysis of quantization table embedded in the JPEG image. But in general, an image of unknown origin is provided for forensic examination without reliable knowledge of post-processing operations applied on it.

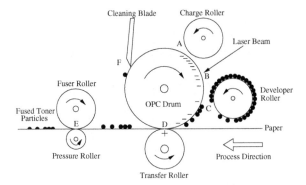

Fig. 3. Diagram of the EP process

And in some cases, revealing the post-processing operations applied on the unknown image is one of the goals of forensic study. Thus, there is need for a general classifier which does not need to know the quality factors of training and testing images.

To design a robust general classifier, the TIFF images compressed at three quality factors are grouped together and 80 randomly chosen images from each scanner class are used for training the classifier. The remaining $(11 * 108 * 3 - 10 \times 80 =)$ 2764 images are used for testing the classifier. The bar graph shown in Fig. 2 shows the classification accuracies of general multi-class classifier. The proposed algorithm has an average classification accuracy of 92.3% among ten scanner models. The proposed scheme gives high classification accuracy even without knowing the JPEG quality factors of training or testing images.

3   Characterization of Printers

Techniques that use the print quality defect known as *banding* in electrophotographic (EP) printers as an intrinsic signature to identify the model and manufacturer of the printer have been previously reported in [9, 10]. We showed that different printers have different sets of *banding frequencies* that are dependent upon brand and model. This feature is relatively easy to estimate from documents with large midtone regions. However, it is difficult to estimate the banding frequencies from text. The reason for this is that the banding feature is present in only the process direction and in printed areas. The text acts as a high energy noise source upon which the low energy banding signal is added.

One solution for identifying intrinsic signatures in text, previously reported in [4], is to find a feature or set of features which can be measured over smaller regions of the document such as individual text characters. If the print quality defects are modeled as a texture in the printed areas of the document then texture features can be used to classify the document. These types of features can be more easily estimated over small areas such as inside a text character.

An understanding of the EP (laser) printing process is necessary in order to gain insight into the types of features that can be used to describe these printers. The first thing to note is that in the printed output from any printer there exist defects caused by electromechanical fluctuations or imperfections in the print mechanism. Because these "print quality defects" are directly related to the printer mechanism, they can also be viewed as an intrinsic signature of the printer. The major components of these intrinsic signature are stable over time and independent of the consumables in the printer.

Figure 3 shows a side view of a typical EP printer. The print process has six steps. The first step is to uniformly charge an optical photoconductor (OPC) drum. Next a laser scans the drum and discharges specific locations on the drum. The discharged locations on the drum attract toner particles which are then attracted to the paper which has an opposite charge. Next the paper with the toner particles on it passes through a fuser and pressure roller which melt and permanently affix the toner to the paper. Finally a blade or brush cleans any excess toner from the OPC drum.

In EP printing, some causes of the artifacts in the printed output are fluctuations in the angular velocity of the OPC drum, gear eccentricity, gear backlash, and polygon mirror wobble. These imperfections in the printer are directly tied to the electromechanical properties of the printer and create corresponding fluctuations in the developed toner on the printed page. The fluctuations in the developed toner can be modeled as a texture. Since the mechanical properties which contribute the most to these fluctuations, such as gear ratios, do not change over time, they can be used reliably to intrinsically identify a printer. In the following sections, two techniques for identifying the printer which created a document are described.

3.1   Halftone Images

In EP printing, the major artifact in the printed output is *banding* which is defined as those artifacts that are due to quasiperiodic fluctuations in process direction parameters in the printer. These are primarily due to fluctuations in the angular velocity of the OPC drum and result in non-uniform scan line spacing. This causes a corresponding fluctuation in developed toner on the printed page. The appearance of the banding artifact is alternating light and dark bands perpendicular to the process direction (the direction the paper passes through the printer). The main cause of banding is electromechanical fluctuations in the printer mechanism, mostly from gear backlash. Because these fluctuations are related to the gearing, the *banding frequencies* present in the printed page directly reflect mechanical properties of the printer.

To estimate the banding frequencies of an EP printer, test pages with midtone graylevel patches created using a line fill pattern were printed and analyzed. These patterns were printed on a set of EP printers and then each patttern was scanned at 2400dpi. Each scanned image, $img(i, j)$, was then projected horizontally to produce $proj(i) = \sum_j img(i, j)$ shown in Fig. 4a. Fourier analysis of the projections was then obtained such as in Fig. 4 which shows spikes at 132 cycles/inch and 150 cycles/inch. Table 1 shows a list of printers and their principle banding frequencies as found by this method.

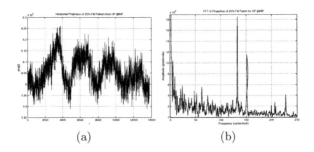

(a)   (b)

Fig. 4. (a) Projection of 25% fill pattern from HP LaserJet. (b) FFT of the projection showing peaks at 132 and 150 cycles/inch.

Table 1. Banding frequencies

Printer Model	Principal Banding Frequencies (cycles/inch)
HP LaserJet 5MP	37, 74
HP LaserJet 6MP	132, 150
HP LaserJet 1000	27, 69
HP LaserJet 1200	69
HP LaserJet 4050	51, 100
Samsung ML-1450	16, 32, 100, 106

Detection and measurement of the banding signal in documents with large midtone regions, such as those with graphic art, can easily be done using methods similar to that used to produce Tab. 1.

3.2   Forensic Characterization of Printed Text

Detection of the banding signal in text is difficult because the power of the banding signal is small with respect to the text, and because only a limited number of cycles of the banding signal can be captured within the height of one text character. Instead, all the print quality defects, including the banding, are lumped together and considered a texture in the printed regions of the document. Features are then extracted from this texture to be used as an intrinsic signature of the printer.

Features are extracted from individual printed characters, in particular the letter "e"s in a document. The reason for this is that "e" is the most frequently occurring character in the English language. Each character is very small, about 180x160 pixels and is non-convex, so it is difficult to perform any meaningful filtering operations in either the pixel or transform domain if the area of interest is only the printed region of each character. To model the texture in the printed regions, graylevel co-occurrence texture features, as well as two pixel based features are used [4].

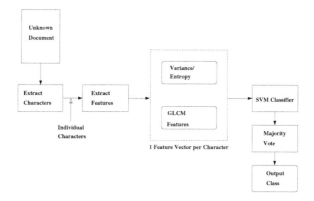

Fig. 5. Diagram of printer identification

Graylevel co-occurrence texture features assume that the texture information in an image is contained in the overall spatial relationships among the pixels in the image. This is done by first generating the Graylevel Co-occurrence Matrix (GLCM). This is an estimate of the second order probability density function of the pixels in the image. The features are statistics obtained from the GLCM as described in [4].

Figure 5 shows the block diagram of the printer identification scheme for text documents proposed in [4]. Given a document with an unknown source, referred to as the *unknown document*, this process can be used to identify the printer that created it. For testing purposes, the Forensic Monkey Text Generator (FMTG) described in [9] is used to create random documents with known statistics to be classified.

The first step is to scan the document at 2400 dpi with 8 bits/pixel (grayscale). Next, all the letter "e"s in the document are extracted. A set of features are extracted from each character forming a feature vector for each letter "e" in the document. Each feature vector is then classified individually using a support vector machine (SVM) classifier.

The SVM classifier is trained with 5000 known feature vectors. The training set is made up of 500 feature vectors from each of 10 printers. Each of these feature vectors generated are independent of one another.

Let Ψ be the set of all printers $\{\alpha_1, \alpha_2, \cdots, \alpha_n\}$. 10 printers are used in our work: { {Brother HL-1440, 1200dpi}, {HP lj4050, 600dpi}, {HP lj1000, 600dpi}, {HP lj1200, 600dpi}, {HP lj5M, 600dpi}, {HP lj6MP, 600dpi}, {Minolta 1250W, 1200dpi}, {Okidata 14e, 600dpi}, {Samsung ML-1430, 600dpi} }. For any $\phi \epsilon \Psi$, let $c(\phi)$ be the number of "e"s classified as being printed by printer ϕ. The final classification is decided by choosing ϕ such that $c(\phi)$ is maximum. In other words, a majority vote is performed on the resulting classifications from the SVM classifier.

Using this process with the GLCM feature set, a high classification rate can be achieved between the 10 printers. A classification matrix showing these results using 300 letter "e"s from 12 point Times documents is shown in Fig. 6. Each

train\test	lj5m	lj6mp	lj1000	lj1200	E320	ml1430	ml1450	hl1440	1250w	14e	Output class
lj5m	296	2	0	1	0	1	0	0	0	0	lj5m
lj6mp	1	256	6	0	17	0	0	15	5	0	lj6mp
lj1000	2	2	284	12	0	0	0	0	0	0	lj1000
lj1200	7	2	2	289	0	0	0	0	0	0	lj1200
E320	0	0	0	0	300	0	0	0	0	0	E320
ml1430	1	0	0	0	0	299	0	0	0	0	ml1430
ml1450	0	0	0	0	0	0	300	0	0	0	ml1450
hl1440	0	28	0	0	0	5	2	259	6	0	hl1440
1250w	0	0	0	0	0	0	0	3	292	5	1250w
14e	0	0	0	0	0	0	0	17	67	216	14e

Fig. 6. Classification results using 300 "e"s from 12 point Times text documents

Fig. 7. Sensor forensics web site at Purdue University

entry of the matrix is the number of "e"s out of the 300 in the test document that were classified as the printer listed at the heading of its column.

In [11], the performance of this printer identification technique was tested for other font sizes, font types, paper types, and age difference between training and testing data sets. The classification results in these cases remains near 90% except for the case where the training data is older than the testing data, in which case the classification rate is near 70%.

4   Conclusion

Forensic characterization of devices is important in many situations today and will continue to be important for many more devices in the future. We have presented an overview of current characterization techniques for scanners and printers. For images scanned at low resolutions such as 200 DPI, it is possible to successfully identify the scanner model, and groups scanners of the same make and model as a single class (Figure 2). An average classification accuracy of 99.9% is obtained among eleven scanners of 10 different models.. The proposed scheme performs well even with images that have undergone JPEG compression with low quality factors. The design of suitable noise features and use of a denoising filter bank which can capture different kinds of scanning noises results in consistently high classification accuracy of the proposed scheme. Likewise, accurate printer identification from printed documents has been shown to be possible using simple image processing and classification techniques. These techniques are shown to be robust against a variety of variables such as document or printer age and paper type, which might be encountered in an actual forensic examination. All of our work can be found at our web site (Fig. 7), `http://sensor-forensics.org/`.

References

1. Khanna, N., Mikkilineni, A.K., Martone, A.F., Ali, G.N., Chiu, G.T.C., Allebach, J.P., Delp, E.J.: A survey of forensic characterization methods for physical devices. Digital Investigation 3, 17–28 (2006)
2. Khanna, N., Mikkilineni, A.K., Chiu, G.T.C., Allebach, J.P., Delp, E.J.: Forensic classification of imaging sensor types. In: Proceedings of the SPIE International Conference on Security, Steganography, and Watermarking of Multimedia Contents IX. SPIE, vol. 6505, p. 65050U (2007)
3. Khanna, N., Chiu, G.T., Allebach, J.P., Delp, E.J.: Forensic techniques for classifying scanner, computer generated and digital camera images. In: Proceedings of the 2008 IEEE International Conference on Acoustics, Speech, and Signal Processing, Las Vegas, NV (March 2008)
4. Mikkilineni, A.K., Chiang, P.J., Ali, G.N., Chiu, G.T.C., Allebach, J.P., Delp, E.J.: Printer identification based on graylevel co-occurrence features for security and forensic applications. In: Proceedings of the SPIE International Conference on Security, Steganography, and Watermarking of Multimedia Contents VII, San Jose, CA, March, vol. 5681, pp. 430–440 (2005)
5. Lukas, J., Fridrich, J.J., Goljan, M.: Digital camera identification from sensor pattern noise. IEEE Transactions on Information Forensics and Security 1(2), 205–214 (2006)
6. Chen, M., Fridrich, J., Goljan, M., Lukas, J.: Determining image origin and integrity using sensor noise. IEEE Transactions on Information Forensics and Security 3(1), 74–90 (2008)
7. Khanna, N., Mikkilineni, A.K., Chiu, G.T.C., Allebach, J.P., Delp, E.J.: Scanner identification using sensor pattern noise. In: Proceedings of the SPIE International Conference on Security, Steganography, and Watermarking of Multimedia Contents IX. SPIE, vol. 6505, p. 65051K (2007)
8. Foi, A., Katkovnik, V., Egiazarian, K., Astola, J.: A novel local polynomial estimator based on directional multiscale optimizations. In: Proceedings of the 6th IMA Int. Conf. Math. in Signal Processing, vol. 5685, pp. 79–82 (2004)
9. Mikkilineni, A.K., Ali, G.N., Chiang, P.J., Chiu, G.T., Allebach, J.P., Delp, E.J.: Signature-embedding in printed documents for security and forensic applications. In: Proceedings of the SPIE International Conference on Security, Steganography, and Watermarking of Multimedia Contents VI, San Jose, CA, January, vol. 5306, pp. 455–466 (2004)

10. Ali, G.N., Chiang, P.J., Mikkilineni, A.K., Chiu, G.T.C., Delp, E.J., Allebach, J.P.: Application of principal components analysis and gaussian mixture models to printer identification. In: Proceedings of the IS&T's NIP20: International Conference on Digital Printing Technologies, Salt Lake City, UT, October/November, vol. 20, pp. 301–305 (2004)
11. Mikkilineni, A.K., Arslan, O., Chiang, P.J., Kumontoy, R.M., Allebach, J.P., Chiu, G.T.C., Delp, E.J.: Printer forensics using svm techniques. In: Proceedings of the IS&T's NIP21: International Conference on Digital Printing Technologies, Baltimore, MD, October, vol. 21, pp. 223–226 (2005)

Evaluation of Graylevel-Features for Printing Technique Classification in High-Throughput Document Management Systems

Christian Schulze, Marco Schreyer, Armin Stahl, and Thomas M. Breuel

German Research Center for Artificial Intelligence (DFKI),
University of Kaiserslautern,
67663 Kaiserslautern, Germany
{christian.schulze,marco.schreyer,armin.stahl}@dfki.de,
tmb@informatik.uni-kl.de
http://www.iupr.org

Abstract. The detection of altered or forged documents is an important tool in large scale office automation. Printing technique examination can therefore be a valuable source of information to determine a questioned documents authenticity. A study of graylevel features for high throughput printing technique recognition was undertaken. The evaluation included printouts generated by 49 different laser and 13 different inkjet printers. Furthermore, the extracted document features were classified using three different machine learning approaches. We were able to show that, under the given constraints of high-throughput systems, it is possible to determine the printing technique used to create a document.

Keywords: feature evaluation, printing technique classification, counterfight detection, questioned document, document forensic, document management.

1 Introduction

As with many new technologies, the opportunity to create printed documents in high quality has resulted in a more extensive usage of these technologies. However this progress, as more and more applicable, is not only used for legitmate purposes but also for illegal activities. With the digital imaging technique available today, it is simple to create forgeries or altered documents within short timeframes. Recent cases reported to the American Society of Questioned Document Examiners (ASQDE) reveal the increasing involvement of modern printing technologies in the production of counterfeited banknotes[1,2] and forged documents[3,4].

In particular within large companies and governmental organizations where paperless processing is aimed, many incoming documents and invoices are handled by large-scale automatic document management systems (DMS). Especially in the case of banks, insurances and auditing companies, processing several thousand documents each day, there is a major need for intelligent methods to determine if the processed documents are genuine or not. Observing the high number

S.N. Srihari and K. Franke (Eds.): IWCF 2008, LNCS 5158, pp. 35–46, 2008.

of processed bills being related to payments and assuming that only a low percentage of these are forged or manipulated, it is easy to imagine that quite some disprofit could be prevented with the use of an authenticity verification system.

The examination of questioned documents usually progresses from the general to the specific[5]. It is a common practice for document examiners to step through their examinations attempting to first determine document class characteristics. Therefore important insights in the examination process can be obtained by answering the question: How the document at hand was created? The ability to investigate documents for consistence in printing technology can be a first useful observation, deciding if the given document is genuine. Furthermore, detecting if specific document regions have been printed with the same non impact printing technique, is an essential piece of information for making the decision.

Therefore, within this paper the performance of common textural and edge based graylevel features developed for digital printing technique recognition was evaluated in a real world scenario. The evaluation was carried out with respect to scan resolution constraints that commonly apply to high throughput scanning systems being used for DMS. The goal of the extensive investigation was to examine the tested features for their applicability to low resolution scans. For this study printouts generated by 49 different laser and 13 different inkjet printers have were evaluated. These printouts were based on a template document whose layout can usually be found in typical office environments. Aspects like the paper quality and ink type used, as well as the effect of document aging have not been taken into account, since the receiver of a document usually has no or only little influence on those details.

1.1 Related Work

Forensic document examiners are confronted on a daily basis with questions like by whom or what device a document was created, what changes have occurred since its original production, and is the document as old as it purports to be[5]. Therefore a variety of sophisticated methods and techniques have been developed since the prominent article [6] published by Albert S. Osborn and Albert D. Osborn in 1941.[1] The textbooks of Hilton[7], Ellen[8], Nickell[9], Kelly and Lindblom[5] offer excellent overviews of the state of the art in the techniques applied to questioned documents by forensic document examiners. These techniques can be divided into destructive and non destructive analysis determining physical and chemical document features.

Although the fact that the use of digital imaging techniques in the forensic examination of documents is relatively new, recent publications show promising and interesting methods in terms of discriminating non impact printing techniques.

---

[1] "A document may have any one of twenty or more different defects that are not seen until they are looked for. Some of these things are obvious when pointed out, while others to be seen and correctly interpreted must be explained and illustrated", by Albert S. Osbornand Albert D. Osborn, co-founders of the American Society of Questioned Document Examiners published the year before the formal founding.

Therefore, a valuable source of information in the determination of a documents underlying printing technology can be gained by an assessment of the print quality. Oliver et al.[10] outlined several print quality metrics including line width, raggedness and over spray, dot roundness, perimeter and number of satellite drops. It is intuitively clear that for an evaluation of these metrics a high resolution scan of the document is inescapable. Another method proposed by Mikkilineni et al.[11] [12] traces documents according to the printing device by extracting graylevel co-occurrence features from the printed letter "e". But their method is based on 1200*dpi* high resolution scans and consequently also not feasible for high throughput systems.

The systems outlined by Tchan[13] exhibit high similarity to our approach. He captured documents with a camera at low resolution and differentiates printing technologies by measuring edge sharpness, surface roughness and image contrast. However, experimental results so far were shown for documents containing simple squares and circles but have not been tested on office documents.

Caused by the reduction in price of color laser printers in recent years, another dimension is added to the document feature space and is more and more recognized within the forensic science community. In [14] Dasari and Bhagvati demonstrated the capability to determine different printing substrates and therefore printing techniques by evaluating the documents hue component values within the HSV colour space. Another cutting edge approach investigated by Tweedy[15] and Li et al.[2,16] are yellow dotted protection patterns distributed on documents printed by color laser printers that are nearly invisible for the unaided human eye. It was demonstrated that these distinctive dotted patterns are directly related to the serial number and could be used for the identification of a particular laser printer. Nowadays, according to our observation, it can not be assumed that documents handled via high throughput systems, are exclusively printed by color laser printers. Therefore, this approach is not suitable for our purposes.

The physical characteristics of printing devices can beside the printing technology also leave distinctive fingerprints on printed documents. As recently shown by Akao et al. [17,18] the investigation of spur gears, holding and passing the paper through the printing device, can also be used to link questioned documents to suspected printers. Therefore the pitch and mutual distance of spur marks on documents was compared to already known printing devices. However this approach is also not applicable in our scenario since knowledge about spur mark distances of different devices is necessary to perform comparisons.

Judging from the literature we were able to review, so far no proper evaluation in real world scenarios of the proposed graylevel features is currently available.

2   Experiments

The evaluation described in the following, was based on four graylevel features as proposed by Lampert et al.[19] but was also covering three additional features originating from the work of Qu[20]. Within this chapter first an outline of

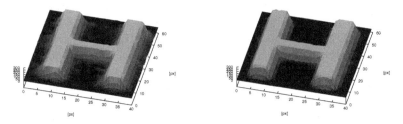

Fig. 1. Surface plots of (l) inkjet and (r) laser printed letter H, scanned with a resolution of 400*dpi*. Notice the sharper edges and cleaner surrounding of the laser printed character (r).

the so far unpublished features of Qu will be presented. In a second step the experimental setup of the evaluation will be explained in detail.

2.1   Features

A close look at the printed characters in Fig. 1 reveals the difference of printing signatures caused by specific printing techniques used in the creation of the document. As immediately obvious, laser and inkjet printed characters can be distinguished according to their differing edge sharpness and satellite droplets of ink. Furthermore, measuring the uniformity and homogenity of ink or toner substrate on printed areas is also a valuable feature in the detection of the used printing technique. The features of Lampert et al.[19] and Qu[20] were explicitly designed to elaborate on this observations. In the following the features proposed by Qu are explained in greater detail:

– **Perimeter Based Edge Roughness**
  An approach for measuring the roughness of a character is to compare the perimeter difference of a binarized and a smoothed binarized image. For the binarization the first valley next to the lowest gray level found in the histogram of the original image is chosen as global threshhold. A character image binarized with threshhold T gives the perimeter p_b. After applying a smoothing with a median filter, the smoothed perimeter p_s can be obtained. The perimeter based edge roughness is then calculated as follows:

$$R_{PBE} = \frac{p_b - p_s}{p_s} \tag{1}$$

– **Distance Map Based Edge Roughness**
  Instead of comparing simply the perimeter values of the binarized image I_b and its smoothed version I_s, this feature relates edge pixel locations via distance mapping. The distance map is initialized with the values taken from the smoothed binary image. Propagating the distances fills all entries of the distance map with the minimal distance to the nearest edge pixels of I_s.

$$DIST = \min\{d|d = \sqrt{(x-m)^2 + (y-n)^2}\}, \tag{2}$$

with $(x, y) \in I_b$ and $(m, n) \in I_s$. This information can be transformed into a distance histogram, where *mean, sample standard deviation, maximal and relative distance* can be calculated to form a feature vector with a relative distance defined as:

$$DIST_{rel} = \frac{\sum\limits_{x \in Edge} dist_{map}(x)}{|Edge|} \tag{3}$$

with *Edge* the set of edge pixels, and the maximal distance:

$$DIST_{max} = \max_{d \in dist_{map}} \{d - \overline{dist_{map}}\} \tag{4}$$

- **Gray Value Distribution on Printed Area**
  As stated above, the differences in the uniformity of ink or toner coverage within printed regions can be used for the determination of the printing technique. To do this, a mask for the printed area is constructed by a variant of Min Max threshholding, only applied to regions containing black pixels. Afterwards a gray value histogram is extracted from the masked image. The coefficients a,b from a regression line, used to characterize the histogram, are used as the feature values.

$$b = \frac{\sum\limits_{i=0}^{n}(x_i - \overline{x})(y_i - \overline{y})}{\sum\limits_{i=0}^{n}(x_i - \overline{x})^2} \tag{5}$$

$$a = \overline{y} - b\overline{x} \tag{6}$$

with

$$\overline{x} = \sum_{i=0}^{n} x_i, \quad \overline{y} = \sum_{i=0}^{n} y_i, \quad n = 255$$

2.2 Experimental Setup

In a first step, well known *document image databases*[2] were evaluated regarding their usability for printing technique classification. Unfortunately, none of them was providing an annotation scheme of the printing technique used for document creation. Therefore the necessity emerged to create our own document-database, annotated with needed *ground truth* information.

An important aspect in terms of ground truth generation is the selection or creation of a suited test-document. In german speaking countries a document called the 'Grauert' letter, implementing the DIN-ISO 10561 standard, is used for the test of printing devices. The 'Grünert' letter[3], which is derived from this

---

[2] UW English Document Image Database I - III, Medical Article Records System (MARS), MediaTeam Oulu Document Database, Google 1000 Books Project.

[3] Exhibiting the following characteristics, fonttype *'Courier New'* normal, fontsize *12 pt*, lineheight *12 pt*.

document, yields the same results in printer tests. Because of its high similarity in layout and content to regular written business letters, we used the 'Grünert' letter as template in database creation.

The prepared documents, consist out of 49 different laser and 14 different inkjet printouts, typically available in (home) office environments. The variety of device manufacturers exhibits all major printer manufacturers[4].

In a next step all documents were scanned using the 'Fujitsu 4860' high speed scanning device. This scanner is especially designed for high throughput scanning procedures and therefore the maximal scanning resolution is limited to $400dpi$. This is a common constraint for such devices due to reasons concerning processing performance and data storage. Therefore, all documents were scanned at $100dpi$, $200dpi$, $300dpi$, $400dpi$ and stored in the TIFF dataformat to avoid further information loss.

To perform classification on character level at least recognition and extraction of the connected components from the scanned document is necessary. After binarization of the scanned image data using Otsu's method[21], a regional growing algorithm as proposed in [22] was applied for detection. Subsequently, the minimal bounding box rectangle of each detected component was calculated and its content extracted.

For the classification of the extracted components the features outlined above were extracted from every image, giving multidimensional feature vectors for their representation. Each of the extracted features was examined through an exhaustive grid search within the features parameter space, to determine their optimal parameter setup.

According to the "no free lunch" theorem, it is desirable to evaluate classification problems with different classifier types. Therefore, the classification performance on the scanned documents was compared using implementations of the C4.5 decision tree[23], a Multilayer-Perceptron[24] and a Support Vector Machine[25]. Furthermore, the classifier parameters were also optimized in terms of high classification accuracy via corresponding grid searches.

As usual, the data was divided into a training and a test set. The classifiers were trained with features extracted from 6 randomly selected inkjet and laser printouts. For all scanned resolutions a ten-fold cross validation using stratified sampling, to avoid overfitting of the learned model, was performed. Let T be the training set and $c_1, ..., c_N$ the training vectors within T. For performance comparison the accuracy mean of all training runs M as defined in equation (7) was obtained for each training and resolution.

$$accuracy_T = \frac{\sum_j^M \left(\frac{\sum_i^N x_{i,T,cor}}{N_T} * 100 \right)}{M} \tag{7}$$

Where $x_{i,T,cor}$ refers to a correct classified character in the training set T of training run j and N_T specifies the total amount of characters within T.

Within the testing phase the learned model of the corresponding classifiers was applied to 7 inkjet and 14 laser printouts. The test printouts were created

---

[4] Hewlett-Packard, Epson, Canon, Ricoh, etc.

by different printing devices and randomly selected. The performance of the classifiers was obtained in the following manner. Let D be a scanned document image and $c_1, ..., c_N$ the extracted characters of D. To compare classification performance on document level, the accuracy rate as stated in equation (8) was calculated for each test document.

$$accuracy_D = \frac{\sum_i^N x_{i,D,cor}}{N_D} * 100 \tag{8}$$

$x_{i,D,cor}$ refers to a correct classified character in D and N_D specifies the total number of characters within D.

3   Evaluation Results

In the following the evaluation results obtained from the experimental setup as described in Sec.2.2 are presented. Therefore, results of the training and the testing phase for each classifier are discussed.

3.1   Decision Tree Classification

The decision tree classification evaluation was based on postpruned trees. Therefore, the confidence threshold for pruning was set to 25% and the minimum number of instances per leaf was set to 2.

Training results: Fig. 2 shows the classification performance of every single feature as well as for all features in the training phase. It can be observed, that features like *edge dist* reach higher classification accuracy with increasing scan resolution. However, features like *area diff*, achieve high classification accuracy at low resolutions. Interestingly an accuracy rate of nearly 95% at 100*dpi* using all features in combination was reached.

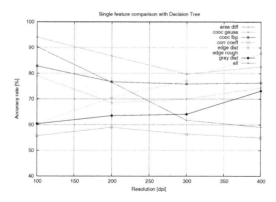

Fig. 2. Accuracy rate for all tested features with a C4.5 decision tree using optimized feature extraction and classification parameters

Testing results: Fig. 3 depicts the appliance of all features to the 21 test documents resulting in the quartile accuracy box plot on the left. Performing a pca on the test set, the three most discriminating features, namely *cooc lbp*, *cooc gauss* and *edge distmap*, could be identified. Classification results for these 3 features taken from the 21 test documents are visible at the right plot of Fig. 3.

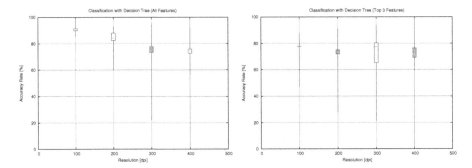

Fig. 3. Box plot of a C4.5 decision tree classification for a combination of all (l) and the 3 most discriminating features as identified by pca (r)

Comparing the classification results, it can be observed that using fewer features leads to a decreasing accuracy for the resolutions. But still $75 - 80\%$ of a documents characters are recognized correctly for all tested resolutions.

3.2   Support Vector Machine (SVM) Classification

Classification experiments using a SVM were based on a radial-basis kernel function using optimized parameters. The parameters for C and γ were obtained by coarse grid searching the SVM parameter space within the intervals $C = [2^{-5}; 2^{15}]$, $\gamma = [2^{-15}; 2^3]$ and for each of the scanned resolutions.

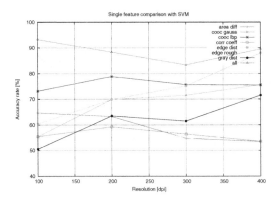

Fig. 4. Accuracy rate for all tested features, classified by a SVM using a rbf kernel and optimized parameters for feature extraction and classification

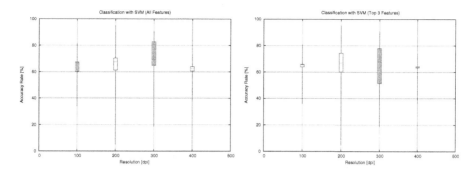

Fig. 5. Box plot of a SVM classification with a combination of all (l) and the 3 most discriminating features as identified by pca (r)

Training results: Overall, the classification results in Fig. 4 are slightly lower than for decision trees considering single features. Also a more constant development of the curves for resolutions $> 200dpi$ can be observed (Fig. 2). Furthermore, a higher classification accuracy at $400dpi$ using all features is achieved.

Testing results: As for the single feature evaluation, the box plots in Fig. 5 show a lower performance for all and the top 3 features. Only $60 - 70\%$ are classified correctly.

3.3   Multi-Layer Perceptron (MLP) Classification

For the classification using a feed forward MLP, the learning rate of the backprop-agation was set to 0.3, which lead to high classification accuracy. Furthermore, within every run the MLP has been trained with 500 training epochs.

Training results: Similar to the SVM, the accuracy rates achieved using a MLP for classification are slightly lower than using the C4.5 decision tree (Fig. 6). Even though some of the features reached higher accuracy rates at low scan resolutions, i.e. *area diff* and *edge rough*, while especially the *edge dist* feature is performing best at $400dpi$. Surprisingly, using all features in combination yields a lower performance than using only the *edge dist* on its own refering to the "ugly duckling" theorem.

Testing results: The classification using all features shows a strong influence of the *edge rough* feature exhibiting only little variance at $100dpi$. Since this feature is not among the top 3 pca features, the accuracy for this resolution drops down a lot (Fig. 7). Regarding results at $400dpi$ the top 3 features improved the accuracy rates significantly in comparison to experiments including all features.

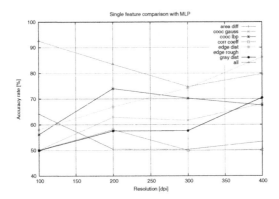

Fig. 6. Accuracy rate for classification of all tested features using a MLP

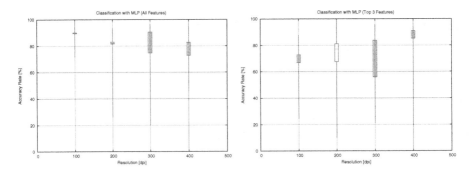

Fig. 7. Box plot of a classification with a multi-layer perceptron for a combination of all (l) and the 3 most discriminating features as identified by pca (r)

4   Conclusion

We have presented a quantitative evaluation of common texture and edge based gray-level features for digital printing technique recognition, under the aspect of usability for high-throughput DMSs. The evaluation indicates that printing technique recognition is possible, even from low resolution scans which are specific to such systems. As the graphs for the single feature evaluation indicate, the examined features perform differently well for the tested resolutions. Therefore the appropriate feature set has to be picked for certain scan resolutions. It was also shown that the classifier used has influence on the feature performance. Furthermore, it could be demonstrated that even classification methods needing only short training, i.e. decision trees, were able to provide high classification accuracy. Due to the constraints of high-throughput document management systems so far only gray-scale scanned documents have been investigated. The examination of color properties of such documents will be part of our future work. Additionally, features capable for examining the differences between a laser print and a copy of a document will be developed. Furthermore the influence of the actually printed shape i.e. even edges versus round edges and its impact on classification accuracy has to be further elaborated.

References

1. Chim, J.L.C., Li, C.K., Poon, N.L., Leung, S.C.: Examination of counterfeit banknotes printed by all-in-one color inkjet printers. Journal of the American Society of Questioned Document Examiners (ASQDE) 7(2), 69–75 (2004)
2. Li, C.K., Leung, S.C.: The identification of color photocopiers: A case study. Journal of the American Society of Questioned Document Examiners (ASQDE) 1(1), 8–11 (1998)
3. Makris, J.D., Krezias, S.A., Athanasopoulou, V.T.: Examination of newspapers. Journal of the American Society of Questioned Document Examiners (ASQDE) 9(2), 71–75 (2006)
4. Parker, J.L.: An instance of inkjet printer identification. Journal of the American Society of Questioned Document Examiners (ASQDE) 5(1), 5–10 (2002)
5. Kelly, J.S., Lindblom, B.S.: Scientfic Examination of Questioned Documents, 2nd edn. CRC Press, Boca Raton (2006)
6. Osborn, A.S., Osborn, A.D.: Questioned documents. Journal of the American Society of Questioned Document Examiners (ASQDE) 5(1), 39–44 (2002)
7. Hilton, O.: Scientific Examination of Questioned Documents, 1st edn. CRC Press, Boca Raton (1993)
8. Ellen, D.: The Scientific Examination of Documents, 2nd edn. Taylor and Francis, Abington (1997)
9. Nickell, J.: Detecting Forgery: Forensic Investigations of Documents, 1st edn. University Press of Kentucky (2005)
10. Oliver, J., Chen, J.: Use of signature analysis to discrimnate digital printing technologies. In: Proceedings of the IS&T's NIP18: International Conference on Digital Printing Technologies, pp. 218–222 (2002)
11. Mikkilineni, A., Ali, G., Chiang, P.J., Chiu, G.C., Allebach, J., Delp, E.: Printer identification based on texture features. In: Proceedings of the IS&T's NIP20: International Conference on Digital Printing Technologies, Salt Lake City, UT, vol. 20, pp. 306–311 (2004)
12. Mikkilineni, A., Ali, G., Chiang, P.J., Chiu, G.C., Allebach, J., Delp, E.: Printer identification based on graylevel co-occurance features for security and forensic applications. In: Proceedings of the SPIE International Conference on Security, Steganography and Watermarking of Multimedia Contents VII, San Jose, CA, vol. 5681, pp. 430–440 (2005)
13. Tchan, J.: The development of an image analysis system that can detect fraudulent alterations made to printed images. In: van Renesse, R.F. (ed.) Optical Security and Counterfeit Deterrence Techniques V, Proceedings of the SPIE, vol. 5310, pp. 151–159 (2004); Presented at the Society of Photo-Optical Instrumentation Engineers (SPIE) Conference, vol. 5310, pp. 151–159 (June 2004)
14. Dasari, H., Bhagvati, C.: Identification of printing process using hsv colour space. In: Narayanan, P.J., Nayar, S.K., Shum, H.-Y. (eds.) ACCV 2006. LNCS, vol. 3852, pp. 692–701. Springer, Heidelberg (2006)
15. Tweedy, J.S.: Class chracteristics of counterfeit protection system codes of color laser copiers. Journal of the American Society of Questioned Document Examiners (ASQDE) 4(2), 53–66 (2001)

16. Li, C.K., Chan, W.C., Cheng, Y.S., Leung, S.C.: The differentation of color laser printers. Journal of the American Society of Questioned Document Examiners (ASQDE) 7(2), 105–109 (2004)

17. Akao, Y., Kobayashi, K., Sugawara, S., Seki, Y.: Discrimnation of inkjet printed counterfeits by spur marks and feature extraction by spatial frequency analysis. In: van Renesse, R.F. (ed.) Optical Security and Counterfeit Deterrence Techniques V, Proceedings of the SPIE, vol. 5310, pp. 151–159 (2004); Presented at the Society of Photo-Optical Instrumentation Engineers (SPIE) Conference, vol. 5310, pp. 129–137 (June 2004)

18. Akao, Y., Kobayashi, K., Seki, Y.: Examination of spur marks found on inkjet-printed documents. Journal of Forensic Science 50(4), 915–923 (2005)

19. Lampert, C.H., Mei, L., Breuel, T.M.: Printing technique classification for document counterfeit detection. In: Wang, Y., Cheung, Y.-m., Liu, H. (eds.) CIS 2006. LNCS (LNAI), vol. 4456. Springer, Heidelberg (2007)

20. Qu, L.: Image based printing technique recognition. Project Thesis (May 2006)

21. Otsu, N.: A threshold selection method from gray level histograms. IEEE Transactions on Systems, Man and Cybernetics (9), 62–66 (1979)

22. Gonzalez, R.C., Woods, R.E.: Digital Image Processing, 3rd edn. Prentice Hall International, Englewood Cliffs (2007)

23. Quinlan, J.R.: C4.5: programs for machine learning. Morgan Kaufmann Publishers Inc, San Francisco (1993)

24. Rumelhart, D., Hinton, G., Williams, R.: Learning Internal Representations by Error Propagation. In: Rumelhart, D.E., McClelland, J.L. (eds.) Parallel Distributed Processing: Explorations in the Microstructure of Cognition. Foundations, vol. 1, pp. 318–362. MIT-Press, Cambridge (1986)

25. Chang, C.C., Lin, C.J.: LIBSVM: a library for support vector machines (2001), http://www.csie.ntu.edu.tw/~cjlin/libsvm

Document Signature Using Intrinsic Features for Counterfeit Detection

Joost van Beusekom[1], Faisal Shafait[2], and Thomas M. Breuel[1,2]

[1] Technical University of Kaiserslautern, Kaiserslautern, Germany
[2] German Research Center for Artificial Intelligence (DFKI), Kaiserslautern,
Germany
{joost.van-beusekom,faisal.shafait,tmb}@dfki.uni-kl.de
http://www.iupr.org

Abstract. Document security does not only play an important role in specific domains e.g. passports, checks and degrees but also in every day documents e.g. bills and vouchers. Using special high-security features for this class of documents is not feasible due to the cost and the complexity of these methods. We present an approach for detecting falsified documents using a document signature obtained from its intrinsic features: bounding boxes of connected components are used as a signature. Using the model signature learned from a set of original bills, our approach can identify documents whose signature significantly differs from the model signature. Our approach uses globally optimal document alignment to build a model signature that can be used to compute the probability of a new document being an original one. Preliminary evaluation shows that the method is able to reliably detect faked documents.

1   Introduction

Document signatures for paper documents, features on a document to prove its originality, have always been a critical issue, even in ancient times, where the number of paper documents was rather limited compared to the number of documents used today. In these days, where modern technologies enable a broad mass of people to easily counterfeit documents and bills, it becomes more and more important to assure that the document comes from the expected source and that it has not been faked or altered. The signet rings from the monarchs that were used to sign the documents in ancient times have nowadays been replaced with all kind of watermarks: specialized paper, holographic images [1], specialized printing techniques [2] and other physical and chemical signatures [3].

All these methods are in one way or another enhancements of the information medium (often this medium is paper) or the printing process, which results in increased costs compared to the use of standard paper and printing devices. Furthermore, in applications involving a high number of different sources of bills or vouchers (many different invoicing parties), these techniques may be impractical due to the high number of invoicing processes that would need to be adapted.

S.N. Srihari and K. Franke (Eds.): IWCF 2008, LNCS 5158, pp. 47–57, 2008.

A typical example of such a use-case involving many different invoicing parties can be found in the tax office: the annual tax declarations are often joined by vouchers from many different invoicing parties. Checking the originality for each voucher is not feasible for tax inspectors. Still it would be important to check for faked bills, as these may be used by tax dodgers to pay less money to the tax office.

In this paper we present a method that allows identifying faked vouchers and bills using a signature obtained from intrinsic features of the documents, namely bounding boxes of connected components. Our method has the advantage, that no extra security features have to be added either to the paper or to the printing process. The main class of forgeries that can be detected by this approach is the imitation of existing bills, which can easily be done by persons capable of using text processing software.

The approach works as follows: observing a number of original bills from one invoice party allows to build a model signature of the non-variable part of a bill. A new bill is then checked against this model signature and if it is significantly different, it is considered as a potentially faked bill. The doubtful bill could then be given to a human operator for further inspection.

The rest of this paper is organized as follows: Section 2 presents our approach in detail and provides an overview of the intrinsic features used in this work. Evaluation and results are shown in Section 3. Section 4 concludes the paper with a short summary and outlook.

2   Description of the Approach

As mentioned in the introduction, we focus on identifying fakes of every day documents, e.g. vouchers and bills. The class of falsification methods we are aiming at is the case of home-made pseudo-copies of the bills, which are created by trying to remake the document using a text processing software. Although the faked documents are often at a first glance quite similar, it is very difficult to obtain exactly the same layout conserving the same spacings. Therefore, we want to identify differences in positions of characters in the static part of the bills. The static part of bills and vouchers are the regions of the page that contain always the same information for one invoicing party, e.g. headers, bank account information, and the contact information of the invoicing party.

The layout of a document can be viewed at different abstraction levels, starting from pixels, over to connected components, and finally lines and paragraphs. We choose connected components, more precisely the bounding boxes of connected components as a suitable description of the document images: connected components are well defined, easy to compute and quite stable, as the scanning process can be influenced to deliver reasonable quality for binarized images of the documents.

After a preprocessing step performing binarization [4] and skew correction [5], first we need to construct document signature. This is done in two steps: first, images of all original bills from an invoicing party are aligned with pixel-accuracy to one reference bill from that party (Section 2.1); second, a signature for this invoicing party is built based on an analysis of variations in positions and sizes of connected components among the aligned bills (Section 2.2). Once a signature is constructed for an invoicing party, the originality of the new bills from that party can be verified by comparing it to the signature (Section 2.3).

We currently focus on the non-variable parts of the documents (e.g. headers, footers, source address and phone number). The distinction between variable and non-variable parts is currently done manually using a bit mask defining which image regions belong to the non-variable part and which do not. The mask can easily be created from one reference document. In future we plan to extract the mask in an automated way by analyzing the bills from one invoicing party using layout analysis methods [6].

2.1   Alignment

The first step in our method is to accurately and robustly align the images. The alignment of two document images aims at identifying the transformation parameters that allow to overlay both images.

Different techniques have been proposed in literature for image registration and alignment. The approaches for general image registration [7] are not well suited for binary document image registration because binary documents lack the color and texture features that are typically used in image registration. Nakai et al. [8] and Liang et al. [9] have proposed image registration techniques for document images, but they handle alignment/registration of the same document under different kinds of distortions. Another way could be to use page frame detection [10] first and then align the page frame of two documents. However, page frame detection is error prone and small errors in the detected page frame will lead to large alignment errors.

In this work, we use the image-matching technique described in [11] for aligning two images from the same invoicing party. This technique is tolerant to changes in the two documents to be matched and hence is a good candidate for use in this scenario. It uses an optimal branch-and-bound search algorithm, called RAST [12] (Recognition by Adaptive Subdivision of Transformation Space). This method allows robust and accurate finding of the globally optimal parameters describing the transformation needed to align both images. Since the RAST algorithm finds the globally optimal alignment of the two images, it is expected that it will align the two images based on their static part.

The quality function used in this case is defined as the number of model points matching an image point under the error bound ϵ.

The RAST algorithm uses a branch-and-bound search for quickly finding a global optimum, which s in our case a maximum for the quality function. The method uses a priority queue containing parameter subspaces in order of their upper bound quality. The highest upper bound quality subspace is divided into two new subspaces, by splitting it into two parts of equal size. For each part, the new upper bound quality is determined and both subspaces are added into the priority queue. These steps are repeated until a stopping criterion is met. In our case the method stops when the size of the remaining parameter sub space is smaller than a given threshold.

For applying RAST, first an initial parameter space (also called transformation space) has to be defined. Let $[tx_{min}, tx_{max}] \times [ty_{min}, ty_{max}] \times [a_{min}, a_{max}] \times [s_{min}, s_{max}]$ be the initial search space, where tx stand for translation in x direction, ty translation in y direction, a for the rotation angle and s for the scale.

Next, computation of the upper bound quality has to be done. Let $B = \{b_1, \ldots, b_N\} \in R^2$ be the set of image points of the scanned image and $M = \{m_1, m_M\} \in R^2$ the set of image points of the synthetic image, also called "model points" (in order to stick to the original notation of the RAST algorithm). For each model point m, a bounding rectangle $G_R(m)$ can be computed using the transformation space to be searched. This rectangle represents the possible positions where a model point m may be transformed to, using all possible transformations from the current transformation subspace. If the distance d, defined as $d = min_{g \in G_R(m), b \in B} ||g - b||$ is less than a threshold ϵ, the quality of the parameter subspace is incremented. A more detailed description of RAST can be found in [12,13].

As image points we choose the centers of connected components, as they are relatively stable and easy to compute. In order to speed up the computation of the upper bound for the quality, a filtering step is added before the branch-and-bound search: to avoid comparing bounding boxes that are not similar at all, Fourier descriptors for the contour of the connected components have been extracted [14], describing the shape of the connected component. In order to be invariant to scale and rotation, the images of the connected components are downscaled to a fixed size and the phase is discarded to obtain rotation invariance for the Fourier Descriptors. For each model point (connected component) only the 50 most similar image points are considered for the quality estimation. The value of 50 was chosen manually and showed to work quite well for standard documents. A more detailed description of the filtering step can be found in our previous work [11].

2.2   Building the Model Signature

We follow a probabilistic approach to compute the probability of a document being original. Let ω_o and ω_f denote the two classes of "original" and "faked" documents respectively. Let X be the observed document image consisting of bounding boxes of connected components $x_1, \ldots, x_n$. The posterior for the observed image to be an original one can then be written as:

$$p(\omega_o|X) = \frac{p(X|\omega_o)p(\omega_o)}{p(X)}$$
$$= \frac{\Pi_{i=1}^n p(x_i|\omega_o)p(\omega_f)}{p(X)}$$
$$= \frac{\Pi_{i=1}^n p(x_i|\omega_o)p(\omega_f)}{p(X|\omega_o)p(\omega_o) + p(X|\omega_f)p(\omega_f)} \quad (1)$$

For Equation 1 we assume independence of the observed connected components, which is not always true, but as documents, bills and vouchers have quite
diverse types of fonts, font sizes and layouts, the assumption is reasonable.

Problems appear for the other parameters: the prior for having a fake has to
be estimated from the dataset. If this not possible, it could be set by the operator
to tune the sensitivity of the method. Another problem is the estimation of the
$p(X|\omega_f)$. Finding a set of faked documents to train on is quite cumbersome and
not feasible in practice.

Therefore, instead of using the Bayesian view, we follow the classical frequen-
tist view of probabilities. We choose to use only $p(X|\omega_o)$. This can be used to
determine the probability of a document being original. The value for a document differing to much from the ones from the training set is a strong hint that
it may be faked.

The next step is to model $p(x_i|\omega_o)$, the probability of observing a given con-
nected component given the fact that the document is original. A connected
component is defined by four parameters x_l, y_l, x_h, y_h defining the lower left and
upper right corners of the bounding box of the connected component.

To avoid modelling the probability in the four dimensional space, we do an
implicit clustering step allowing to reduce the dimensionality of the resulting
histogram: two components are considered being the same, when their normalized overlap is greater than a given threshold T. The normalized overlap is
defined by:

$$D_{ov}(\mathbf{x}_i, \mathbf{x}_j) = 1 - \frac{2 \times \mathrm{Ov}(\mathbf{x}_i, \mathbf{x}_j)}{\mathrm{area}(\mathbf{x}_i) + \mathrm{area}(\mathbf{x}_j)}$$

where $\mathrm{Ov}(\mathbf{x}_i, \mathbf{x}_j)$ is the number of overlapping pixels of both connected components and $area(c)$ is the number of pixels of connected component x.

We obtain a 2D histogram where the bins represent the positions (the sizes
of connected components are included implicitly) and the height the number of
connected components of similar size at similar positions.

The procedure to extract this histogram from the training images is as follows: first, all images are aligned using RAST, so that the coordinate systems of
all document images have same origin and unit vectors. This is needed to allow
comparison between the positions of connected components of the different documents. The set of scanned original documents from the same source is denoted
as $O = \{B_1, \ldots B_n\}$. One document is taken as reference document, e.g. let B_1
be the reference document. For this document the mask defining what regions
should be considered as fix is manually created.

For each document, the set of connected components is computed. Let us denote the set of all connected components of all documents as $X = \{x_1, \ldots, x_m\}$.
Let M be the bins of the sparse 2D histogram. These bins will be represented
by connected components together with number of samples in the bin. We start
with an empty histogram. Now for each connected component in X, it is checked
if there is a connected component in the model for which the normalized overlapping area is greater than a certain threshold T. Without much parameter tuning
$T = 0.8$ showed to work fine. If this is the case, the counter for the number of

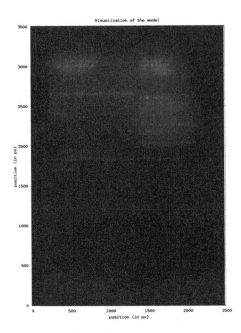

Fig. 1. Visualization of the model. The red regions show regions with stable connected components. The more bluish regions are regions where positions of the connected components are more likely to vary.

samples in the bin is increased. If no such component is found, the given component is added to the histogram as a new bin. The bin sizes are then normalized by the number of all components.

The resulting 2D histogram defines the probability of a connected component of a given size being at a certain position. A simplified visualization of the model can be found in Figure 1.

As the connected components depend on print an scan quality, the question of robustness against merging and breaking connected components arises. As the scanning process can be optimized by the operator of the system, the remaining source of merged and broken components is the invoice generation process. It may happen that the printer of the person creating the bill is low on ink or the paper was changed which could lead to more ink smearing. These problems will result in merged and broken connected components and thus the risk of a false positive will increase. But as these cases should be rare, the cost of sending these to an operator will be reasonably low.

2.3   Checking a New Document

To check if a new document is likely to be an original one or not, the scanned version of the document is aligned to the reference document of the model set B_1. Then, the connected components are extracted. The probability of a connected component to be part of the model is computed using the histogram obtained from the model.

For badly faked bills it may happen, that the alignment will fail. This is no problem as the obtained probability will then be even lower and the faked bill will be reported as falsification.

The probability of a document being original is obtained by:

$$p(X|\omega_o) = \Pi_{i=1}^{n} p(x_i|\omega_o) \tag{2}$$

To decide if this value is likely to be an original document, a threshold value for the probability is defined. This can be set by a human operator. If an automatic setting is needed and if no faked documents are available, the 99% confidence interval of the training set values could be used as a decision rule Under the assumption that the obtained probabilities for the training set are distributed normally, the mean and the variance of the probabilities of the training set can be computed. If the probability of a new bill is less than the mean minus three times the variance, it is classified as a fake.

3   Evaluation and Results

In order to test the performance of our approach, a dataset was needed. As to our best knowledge no public dataset is available containing original and faked

(a)   (b)

Fig. 2. The left image is an "original" document from the dataset. The image to the right represents a sample of the faked documents.

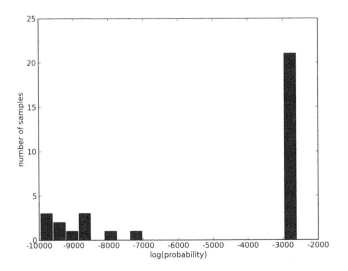

Fig. 3. Histogram of the log-likelihoods. The peak at the right results from the training images. The probabilities of most of the documents are widely different from the ones of the original documents. Only on document had a probability close to the original ones, but still less than all the originals on (around -3000).

documents from one and the same invoice party, we created our own dataset of medical doctor bills as a use case. These sample documents were created by a student using Open Office. Next, we picked randomly one document and gave it to other students. Their task was to copy the document as accurate as possible using the text editor of their choice. The number of original documents is 40, the number of faked documents is 12. An example of an original and a faked document can be found in Figure 2.

The set of original bills was split into a test set and a training set of 20 bills each. The model was trained on 20 original documents. The first document was chosen as reference document, where all the other documents were aligned to. Then the model signature was build. For defining the threshold, the 99% confidence interval has been used, computed on the training set.

Then using the model, the probability for the faked bill of being an original for was computed.

The results of the test are shown in Figure 3. It shows the histogram of the probabilities of all the documents, training set and faked test set, for being original. The peak around the right comes from the original documents in the test set. Using the above mentioned threshold, all the faked documents (12 in total) were correctly classified as fakes. 5 out of the 20 original in the test set were wrongly classified as fakes.

A second test has been done in order to measure the performance of the method on a second falsification scenario: instead of remaking the whole document using a word processor, the forger could just scan an original bill, make some changes using an image editor and print the changed bill. As scanning and

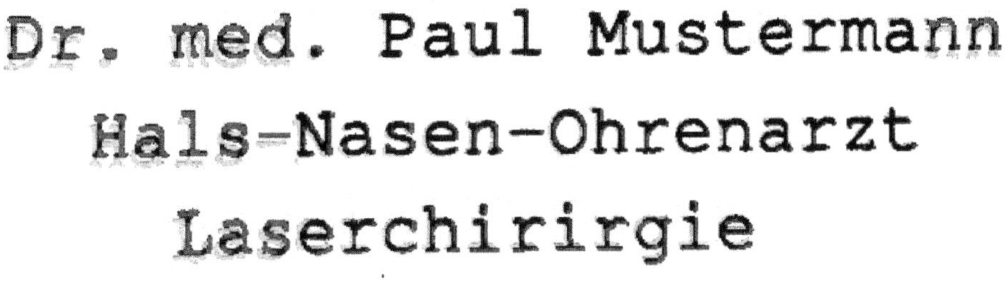

Dr. med. Paul Mustermann
Hals-Nasen-Ohrenarzt
Laserchirirgie

Fig. 4. Example of a distortion induced by copying an original bill. The originally black pixels from the copies bill are painted in blue, the black pixels from the original bill are painted in red. If blue and red pixels overlap, these are painted black. It can be seen that the copying process seems to move blocks up and down: left part the copy is to far down, the middle part fits quite well and the right part is again to far down.

original bills
fakes
copies
-2000
-4000
-6000
-8000
-10000
-12000
0
20
40
60
80

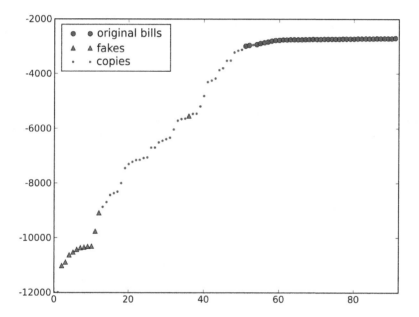

Fig. 5. Plot of the sorted log probabilities (y axis) together with the type of the bill. Circles represent the original bills, triangles the faked bills and dots the copied bills. The x axis represents the plotted bills (92 samples in total).

printing an original bill distorts the bill slightly, our method should be able to detect these cases. An example of such a distortion can be found in Figure 4.

To simulate this scenario we chose 35 bills to train the model. The remaining 5 original bills were copied on different multi function printers (MFP) to simulate effects of scanning and printing the original bill. In total 8 different MFPs were used to obtain 40 copied bills. The copies were then scanned using the same scanner as for the original bills. The threshold is computed in the same manner as done in the test before. Using our approach, all copies were correctly recognized as fakes. From the 5 original bills only one was wrongly classified as faked. The other 4 were correctly classified as original bills, which is equal to an error rate of 2.2%. On the training set, only 2 false positives were registered.

A plot of the sorted log-likelihoods can be found in Figure 5. It can clearly be seen that most originals have high probability of being original, whereas copied bills lie in between the faked ones and the original ones. The two outliers to the left are due to two copies where the toner was nearly empty.

4   Conclusion and Future Work

In this paper we presented a novel approach of document falsification detection using intrinsic document features. Using document alignment a connected component based signature could be computed allowing to estimate the probability of a document to be original.

Our approach was tested on a manually created dataset of doctor bills containing a small number of faked documents. The preliminary results proved that the method works reasonably well. A second test on copies of original doctor bills showed that the approach is even able to detect copies with reasonably high accuracy.

One main conclusion of this test is that it is not easily feasible to exactly counterfeit a bill. Although at a first glance copies and fakes look very similar, more detailed analysis shows, that the small variances are unavoidable due to imperfection of the hardware (MFP in our case). The hypothesis, that would need to be investigated in much more detail is that for bills generated using PCs, exactly faking a bill is not feasible unless the same operating system, the same word processing software and the same printer is used.

This approach could be combined with other intrinsic features, e.g. the features used for printing technique classification [15]. This could be incorporated into the model and allow a more accurate modeling of the invoice party, reducing the risk of missing counterfeits.

One important part of future work is to test the approach more thoroughly. If no data sets with appropriate data can be found, these will have to be generated manually. One weakness currently is the missing modelling of the connected component distribution for faked documents. Furthermore, the method needs to be adapted to work on the whole document and not only on the invariant parts. One example could be the detection of incorrect line spacing in the body part of the document, which could result from belated adding of lines.

References

1. Smith, P.J., O'Doherty, P., Luna, C., McCarthy, S.: Commercial anticounterfeit products using machine vision. In: van Renesse, R.L. (ed.) Optical Security and Counterfeit Deterrence Techniques V. Presented at the Society of Photo-Optical Instrumentation Engineers (SPIE) Conference. Proceedings of the SPIE, vol. 5310, pp. 237–243 (June 2004)

2. Amidror, I.: A new print-based security strategy for the protection of valuable documents and products using moire intensity profiles. In: van Renesse, R.L. (ed.) Optical Security and Counterfeit Deterrence Techniques V. Proceedings of the SPIE, vol. 4677, pp. 89–100 (June 2002)
3. Hampp, N.A., Neebe, M., Juchem, T., Wolperdinger, M., Geiger, M., Schmuck, A.: Multifunctional optical security features based on bacteriorhodopsin. In: van Renesse, R.L. (ed.) Optical Security and Counterfeit Deterrence Techniques V. Proceedings of the SPIE, vol. 5310, pp. 117–124 (June 2004)
4. Bernsen, J.: Dynamic thresholding of gray level images. In: Proc. Int. Conf. on Pattern Recognition, Paris, France, pp. 1251–1255 (1986)
5. Breuel, T.M.: The OCRopus open source OCR system. In: SPIE Document Recognition and Retrieval XV, San Jose, USA, pp. 0F1–0F15 (January 2008)
6. Shafait, F., Keysers, D., Breuel, T.M.: Performance evaluation and benchmarking of six page segmentation algorithms. IEEE Trans. on Pattern Analysis and Machine Intelligence 30(6), 941–954 (2008)
7. Zitova, B., Flusser, J.: Image registration methods: a survey. Image and Vision Computing 21(11), 977–1000 (2003)
8. Nakai, T., Kise, K., Iwamura, M.: A method of annotation extraction from paper documents using alignment based on local arrangements of feature points. In: Int. Conf. on Document Analysis and Recognition, Curitiba, Brazil, pp. 23–27 (September 2007)
9. Liang, J., DeMenthon, D., Doermann, D.: Camera-based document image mosaicing. In: Int. Conf. on Patt. Recog., Hong Kong, China, pp. 476–479 (August 2006)
10. Shafait, F., van Beusekom, J., Keysers, D., Breuel, T.M.: Page frame detection for marginal noise removal from scanned documents. In: Ersbøll, B.K., Pedersen, K.S. (eds.) SCIA 2007. LNCS, vol. 4522, pp. 651–660. Springer, Heidelberg (2007)
11. van Beusekom, J., Shafait, F., Breuel, T.M.: Image-matching for revision detection in printed historical documents. In: Hamprecht, F.A., Schnörr, C., Jähne, B. (eds.) DAGM 2007. LNCS, vol. 4713, pp. 507–516. Springer, Heidelberg (2007)
12. Breuel, T.M.: A practical, globally optimal algorithm for geometric matching under uncertainty. Electronic Notes in Theoretical Computer Science 46, 1–15 (2001)
13. Breuel, T.M.: Implementation techniques for geometric branch-and-bound matching methods. Computer Vision and Image Understanding 90(3), 258–294 (2003)
14. Granlund, G.H.: Fourier Preprocessing for Hand Print Character Recognition. IEEE Trans. on Computers C–21(2), 195–201 (1972)
15. Lampert, C.H., Mei, L., Breuel, T.M.: Printing technique classification for document counterfeit detection. In: Wang, Y., Cheung, Y.-m., Liu, H. (eds.) CIS 2006. LNCS (LNAI), vol. 1, pp. 639–644. Springer, Heidelberg (2007)

Automatic Feature Extraction from 3D Range Images of Skulls[*]

Lucia Ballerini[1], Marcello Calisti[2], Oscar Cordón[1], Sergio Damas[1], and Jose Santamaría[3]

[1] European Centre for Soft Computing, Edf. Científico Tecnológico, Mieres, Spain
{lucia.ballerini,oscar.cordon,sergio.damas}@softcomputing.es
[2] Dpt. of Electronics and Telecommunication, University of Florence, Florence, Italy
marcello.calisti@gmail.com
[3] Dpt. of Computer Science, University of Jaén, Jaén, Spain
jslopez@ujaen.es

Abstract. The extraction of a representative set of features has always been a challenging research topic in image analysis. This interest is even more important when dealing with 3D images. The huge size of these datasets together with the complexity of the tasks where they are needed demand new approaches to the feature extraction problem. The need of an automatic procedure is specially important in many forensic applications, including the reconstruction of 3D models. In this work we propose a new method to automatically extract a set of relevant features from points clouds acquired by a 3D range scanner. We present our results over five views of five skulls, one of them corresponding to a pathological case.

1 Introduction

A laser range scanner is a device that captures the shape of an object and generates a 3D point cloud from the object surface. These scanners have become more affordable and accurate, what makes them suitable for many applications. Consequently, geometric processing of point clouds is becoming increasingly important.

Feature extraction is a primary concern for many geometrical computations and modeling applications. Feature extraction is a well-studied research area in computer vision and medical imaging. However, most of the past research efforts concentrated on data defined in the Euclidean domain, e.g. 2D images and volume data. Feature extraction of surfaces has gained less attention, but it is important in many fields such as range data analysis, where it can be used to support early preprocessing steps including surface reconstruction and adaptive decimation.

---

[*] This work was partially supported by the Spain's Ministerio de Educación y Ciencia (ref. TIN2006-00829) and by the Andalusian Dpto. de Innovación, Ciencia y Empresa (ref. TIC1619), both including EDRF fundings.

S.N. Srihari and K. Franke (Eds.): IWCF 2008, LNCS 5158, pp. 58–69, 2008.

Techniques for feature extraction on mesh- and point-based models have been investigated by several authors. Gumhold et al. [1] use covariance analysis for classification and compute a minimum spanning graph of the resulting feature nodes. Pauly et al. [2] extended this scheme using a multi-scale approach capable of processing noisy data. They modeled the extracted features using *snakes* [3]. Geometric snakes have been used by Lee et al. [4] to extract feature lines in triangle meshes based on normal variation of adjacent triangles. The extraction of *crest lines* has been introduced by Monga et al. [5], and their importance in medical images, especially in skull models, have been emphasized by Subsol et al.[6]. However mesh-based techniques require assumption of connectivity and normals associated with the vertices of the mesh. On the other hand, the accuracy of point-based techniques depends on the sampling quality of the input model. The method we propose aim to overcome such limitations. It automatically extracts a set of relevant features directly from point clouds. We also assume that no attribute information, e.g, normal vectors are given a priori.

The need of an automatic feature extraction method is specially important in forensic anthropology for different applications and, in particular, to obtain 3D models of forensic objects [6,7]. This is the reason that motived us to develop an automatic method able to identify meaningful features in 3D skull models. The work described in the present paper is part of a large forensic project, which final goal is the identification of missing people by means of photographic supra-projection [8]. Several approaches regarding 3D face models have been also presented [9]. They are out of the scope of this paper, as they deal with face recognition that is a completely different problem.

The structure of the paper is as follow. In Section 2 we review some properties of surfaces and the definition of crest lines, that will be later used to compare the results of our automatic method with. Our proposal is presented in Section 3. In Section 4 we apply our method to a set of twenty five 3D range images of skulls. Finally, in Section 5 we present some conclusions and future works.

2   Crest Lines Feature Extraction

This section is devoted to introduce some information that can be derived from the shapes included in the images. In particular, we are interested in the description of the method to extract crest line features that have demonstrated to be suitable for skull modeling [6]. To do so, let us define the iso-intensity surface of a 3D image, which will be called simply the iso-surface in the rest of this paper. For any continuous function $C(x, y, z)$ of $\mathbb{R}^3$, any value I of $\mathbb{R}$ (called the iso-value) defines a continuous, not self-intersecting surface, without hole, which is called the iso-intensity surface of C [5]. A non ambiguous way to define the iso-surface is to consider it as being the surface which separates regions of the space where the intensity of C is greater or equal to I from those regions whose intensity is strictly lower than I. Whether such an iso-surface corresponds to the boundary of the scanned object or not is another problem, that will not be considered in the current contribution. In many cases, such as for medical

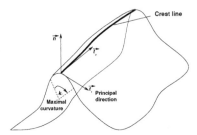

Fig. 1. Differential characteristics of surfaces

images, iso-surface techniques are directly used to segment the organs, for example the bones in CT scans. In other applications, it is necessary to use iso-surface techniques as the final phase of the process to extract the surface in order to ensure that the reconstructed surfaces are continuous, not self-intersecting, and without hole (except of course for the image boundary). Because of those good topological properties, iso-surface techniques are the most widely used methods of segmentation for 3D medical images.

Let us see now some properties of the iso-surfaces (see Figure 1). At each point P of those surfaces, there is an infinite number of curvatures but, for each direction t in the tangent plane at P, there is only one associated curvature k_t. There are two privileged directions of the surface, called the principal directions $(t_1$ and $t_2)$, which correspond to the two extremal values of the curvature: k_1 and k_2. One of these two principal curvatures is maximal in absolute value (let say k_1), and the two principal curvatures and directions suffice to determine any other curvature at point P. These differential values can be used in many different ways to locally characterize the surface. To those values, we can add the extremality criterion e as defined by Monga et al. in [5], which is the directional derivative of the maximal curvature (let say k_1), in the corresponding principal direction (t_1). In fact, the same extremality criterion can be also defined for the other principal direction, and we therefore have two "extremalities" e_1 and e_2. The locations of the zero-crossing of the extremality criterion define lines, which are called ridge lines or crest lines.

Crest lines provide a useful subset of skull data, because they have a very strong anatomical meaning (as pointed out by Subsol et al. [6] they emphasize the mandible, the orbits, the cheekbones or the temples), but their extraction requires the expertise of human intervention. The latter drawback is a big deal for the forensic anthropologists, because they do not have neither the knowledge nor the time to extract them.

3   Feature Extraction Proposal

In order to overcome the drawback pointed out in the previous section, we aim to propose a method that can automatically identify meaningful features. These features should be:

- representative of the skull object we will deal with
- invariant to rigid transformation between adjacent views
- composed by the smallest number of points
- robust to deal with not uniform sampling of points

As said, we want the features to be representative of the object, i.e. the skull. We focus our attention on some regions corresponding to significant anatomical parts: orbits, nasal cavity and cheekbones (see Figure 2 (left)). These regions appear surrounding holes in the surface. Let us formally characterize these regions.

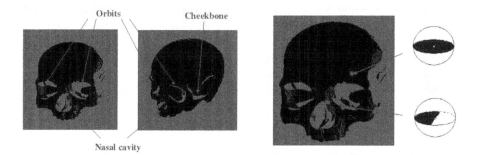

Fig. 2. Left: Two views of a skull. The arrows indicate important anatomical regions we are interested in. Right: The intersection of two spherical neighborhoods with the skull surface: note the different intersection areas.

At each point $P = (p_x, p_y, p_z)$ of those surfaces, we can define a spherical neighborhood N_p whose points $Q = (q_x, q_y, q_z)$ satisfy the condition:

$$\sqrt{(p_x - q_x)^2 + (p_y - q_y)^2 + (p_z - q_z)^2} \leq r \qquad (1)$$

where r is the radius of the neighborhood.

If we consider the skull object, the intersection between the sphere and the surface is different in distinct regions (see Figure 2 (right)). The area of this intersection is smaller when the sphere is located in boundary regions, i.e. in our regions of interest. Hence, we will be able to identify different regions by measuring the size of the intersection area.

We can define a weighted density at each point P_i of the surface S_p of the skull:

$$d_i = \frac{\sum_{j=1}^{n} \| P_i - M_j \|}{4\pi r^3 / 3} \qquad (2)$$

where $M = \{M_1, \ldots M_n\}$ is the set of points of the skull that are inside the sphere centered in P_i with radius r.

Notice that the volume of the sphere in Eq. 2 is a constant value. Hence it can be neglected without affecting the results:

$$w_i = \sum_{j=1}^{n} \| P_i - M_j \| \tag{3}$$

Assuming a roughly uniform sampling of the surface, w_i is directly related to the area of the intersection between the surface and the spherical neighborhood. Thus, w_i will be higher in regions where the intersection area is larger. On the contrary, boundary regions interesting for us will correspond to lower values of w_i. However, the uniform sampling assumption cannot be always guaranteed (see Figure 3). In these case we observed that boundary regions are more crowded than the rest. That is problematic because the w_i values will then be higher in these regions that will become considered as not relevant. To avoid this situation we applied a power-law transformation to each component of w_i to increase the contribution of the farther points while decreasing the contribution of the closer ones:

$$u_i = \sum_{j=1}^{n} \| P_i - M_j \|^5 \tag{4}$$

Hence, we are characterizing every point based on the local neighborhood. Recalling that our regions of interest have smaller intersection area, and consequently contain points having small values of u_i, a suitable threshold on this quantity can identify different areas and therefore segment our surface in two different regions. Points having values of u_i less than the threshold belong to our regions of interest.

In other words, the subset of relevant features can be defined as:

$$S_{p'} = \{p_i \in S_p | u_i \leq \alpha \cdot u_{max}\} \tag{5}$$

where u_{max} is the u_i maximum value and $\alpha \in [0,1]$ is a threshold to identify the relevant regions. The coefficient α can be chosen analyzing the distribution of the u_i values. As said before, the number of points should be enough to represent the important features while being as smallest as possible. Thus, we propose $\alpha = 0.5$

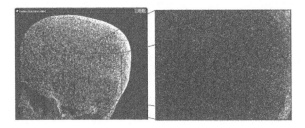

Fig. 3. A skull image and the zoom of a part where the dispersion effect can be observed

Fig. 4. Overview of the proposed method for invariant feature extraction

Figure 4 gives an overview of the proposed method for the extraction of invariant features. Given a point sampled surface $S_p = \{\boldsymbol{p}_i\}$, for each point $\boldsymbol{p}_i$ calculate the value u_i using Eq. 4. Threshold these values to obtain a subset of point $S_{p'}$ representing the searched features. The point sampled surfaces S_p and $S_{p'}$ are the input and output image, respectively.

Our algorithm can also be seen as a segmentation algorithm. Indeed, it performs a partition of the input image into two regions: one contains the feature points and the other contains the remaining points.

4   Applications and Results

Our primary goal in developing the feature extraction algorithm was to identify a subset of points that can improve the registration of multiple views and reconstruct an accurate 3D model of skull objects.

A robust method for 3D range image registration, based on *evolutionary algorithms* [10], have been described in a previous paper [7]. The method consists of a multi-stage approach: in the first stage, the *Scatter Search* algorithm [11] is used for pre-alignment; in the second stage a local optimizer is used for refinement. Typically, in the pre-alignment step only a subset of points is used, while the refinement step is applied to the whole images. In [12], we evaluated two point selection approaches: a semiautomatic and an automatic one. The crest lines are used by the semiautomatic approach, while a subset of randomly chosen points is used by the automatic one. Due to the skull shape, a higher number of points is required by the random sampling to be sure to select points on meaningful anatomical regions and not only in the large flat areas. The semiautomatic outperformed the automatic one, but it required the expertise of human intervention. On the other hand, the automatic approach, due to the higher number of points, required more computation time to achieve a suitable accuracy. The automatic feature extraction method proposed in the present work is an optimal trade-off between having a fully automatic method and low number of points located on meaningful features, that can reduce the computational time. Indeed, our algorithm has been proved to be an efficient aid to the registration tool that we developed for forensic anthropologists [13].

The features extracted by our algorithm can also be used as input by a number of different applications, including the building of anatomical atlas and the identification of craniometric landmarks.

In the following of this section we describe the dataset of forensic objects used for this studies and we present the results of our feature extraction algorithm.

4.1   Input Range Images

The Physical Anthropology Lab at the University of Granada, Spain, provided us with a number of datasets of human skulls acquired by a Konica-Minolta© 3D Lasserscanner VI-910. It should be highlighted that the five skulls considered for this experimental study were chosen by the experts according to several forensic criteria to guarantee a maximal differentiation regarding to skull features. The acquisition process includes noise removal and the use of smoothing filters.

To ease the forensics' work, we have taken into account important factors regarding to the scanning process like time and storage demand. Indeed, we consider a scan every 45°, that is a reasonable trade-off between number of views and overlapping regions. Hence, we deal with a sequence of only eight different views: $0° - 45° - 90° - 135° - 180° - 225° - 270° - 315°$, which supposes a great reduction both in the scanning time and storage requirements. The datasets we will use in our experiments are limited to five of the eight views: $270° - 315° - 0° - 45° - 90°$. The reason is that these views contain the most interesting parts of the skull, i.e. its frontal part. These five views of every skull will provide us twenty five different images for applying our method. Figure 5 shows the five different views of three skulls of our dataset.

4.2   Results

Figure 6 shows the crest lines extracted from the five views shown in Figure 5. Yoshizawa et al.'s proposal [14] was considered to compute the curvature of each point of the surface and to extract the crest lines. We can observe that the crest lines have a strong anatomical significance. Nevertheless, some crest lines appear noisy and filtering is needed to solve this problem. Moreover, the crest lines do not always correspond to the topology expected by anatomists, for example, the orbital crest lines are not closed. Only an a priori model could add this constraint. Furthermore, the main limitation is that their extraction requires human intervention. Instead, our feature extraction method is fully automatic.

Figure 7 shows our extracted features from the five views shown in Figure 5. Based on the skull dimension and its anatomical knowledge, the size of the neighborhood has been fixed to $r = 5mm$. On the other hand, the coefficient $\alpha = 0.475$, chosen as described earlier, proved to be appropriate for our objects and their feature shapes. Our method is run on a PC with an Intel Pentium D820 (2 core 2.8 GHz) processor.

Table 1 summarizes the size (number of surface points) of the forensic range images of the considered skulls. The original size and the size after the application of our extraction procedure are reported. The lower number of points resulting in $Skull_1$ are due both to the pathological alterations of this case and its irregular sampling already shown in Figure 3. However, the good performance of the method also in this case confirm its robustness.

The proposed method is fully automatic, easy to implement and fast from the computational point of view. The values of the parameters needed by the method (r and α) have been selected empirically in our experiments and do not

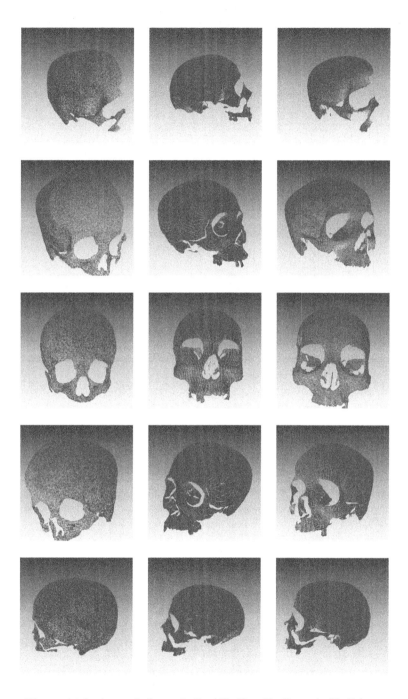

Fig. 5. Five partial views of three skulls ($Skull_1$, $Skull_3$ and $Skull_5$) acquired at $270°, 315°, 0°, 45°$, and $90°$, respectively

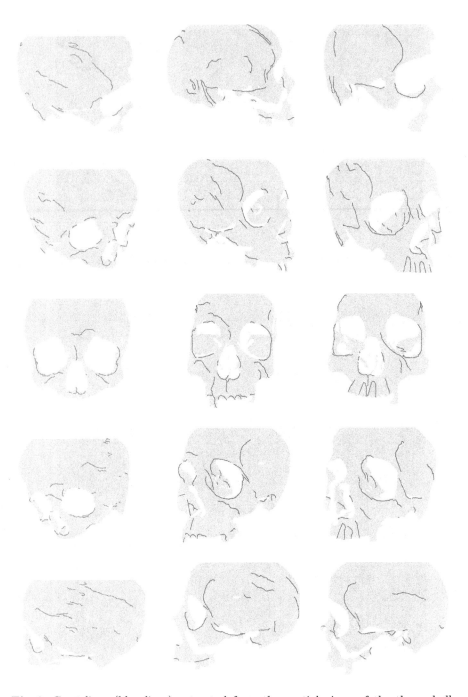

Fig. 6. Crest lines (blue lines) extracted from the partial views of the three skulls shown in Figure 5

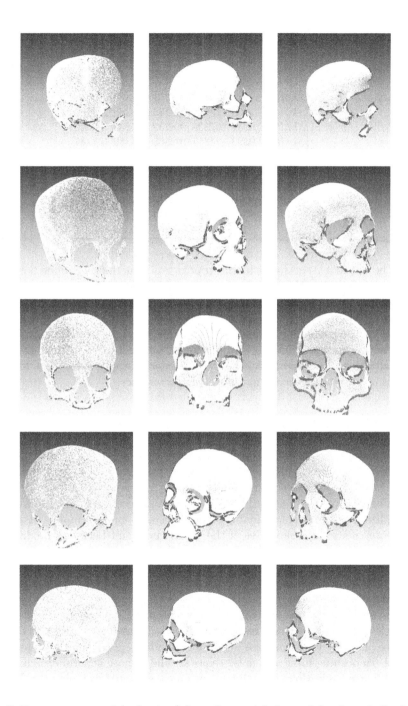

Fig. 7. Features extracted (red points) from the partial views of the three skulls shown in Figure 5

Table 1. Size of the range images of the considered objects in their original conditions and after the automatic feature extraction process

		Views/Images				
		270°	315°	0°	45°	90°
Original	$Skull_1$	109936	76794	68751	91590	104441
	$Skull_2$	121605	116617	98139	118388	128163
	$Skull_3$	116937	107336	88732	111834	123445
	$Skull_4$	129393	124317	102565	125859	137181
	$Skull_5$	110837	102773	83124	101562	110313
Features	$Skull_1$	5199	915	2901	2948	1655
	$Skull_2$	7304	10347	11106	12676	11143
	$Skull_3$	9023	10745	8318	12265	10361
	$Skull_4$	8593	11020	14844	12285	10025
	$Skull_5$	9419	9852	10764	10308	9175

need to be tuned by the forensic anthropologists. Our method does not need any mesh information. Moreover, it does not require calculation of curvature values, unlike the crest line extraction. Hence it is faster and simpler. Besides, it is accurate enough for our scope.

As said at the beginning of this section, our primary goal was to improve the registration of multiple views of skulls. Hence, our features are skull-oriented. A generalization of the method is needed in order to apply it to other forensic objects that present different anatomical characteristics.

5   Concluding Remarks

We have presented a new approach for feature extraction from range images, represented as point clouds. The proposed approach overcomes the trade-off between having a fully automatic method and a low number of points located on meaningful features. Our method can detect regions close to boundaries, and also localize small and sharp features where usually undersampling happens. These features are: representative of the object, invariant between the different views of the object and composed by the smallest number of points. Moreover, the method demonstrated its robustness when dealing with not uniform sampled surfaces. These properties make our approach an excellent preprocessing step for the registration algorithm.

We have shown our results over a number of skulls, one of them presenting pathological alterations. Good results have been obtained also with this case.

We are planning to extend this study by including an automatic preprocessing stage (smoothing filter and noise removal) and a more general design of the invariant feature selection. Actually, our primary goal in developing the feature extraction method was to find some invariant features to improve our skull registration. Hence, our features are skull-oriented. However, the method can be easily extended for general purpose and applied to other objects.

Acknowledgments

We want to acknowledge all the team of the Physical Anthropology Lab at the University of Granada (headed by Dr. Botella and Dr. Aleman) for their support during the data acquisition and validation process.

References

1. Gumhold, S., Wang, X., MacLeon, R.: Feature extraction from point clouds. In: Proc. 10th International Meshing Roundtable, pp. 293–305 (2001)
2. Pauly, M., Keiser, R., Gross, M.: Multi-scale feature extraction on point-sampled surfaces. Computer Graphics Forum 20(3), 385–392 (2001)
3. Kass, M., Witkin, A., Terzopoulos, D.: Snakes: Active contour models. International Journal of Computer Vision 1(4), 321–331 (1988)
4. Lee, Y., Lee, S.: Geometric snakes for triangular meshes. Computer Graphics Forum 21(3), 229–238 (2002)
5. Monga, O., Benayoun, S., Faugeras, O.D.: Using partial derivatives of 3D images to extract typical surface features. In: Proceedings of the IEEE Computer Vision and Pattern Recognition (CVPR 1992), Illinois (USA), pp. 354–359 (1992)
6. Subsol, G., Thirion, J.P., Ayache, N.: A scheme for automatically building three-dimensional morphometric anatomical atlases: application to a skull atlas. Medical Image Analysis 2(1), 37–60 (1998)
7. Santamaría, J., Cordón, O., Damas, S., Alemán, I., Botella, M.: A scatter search-based technique for pair-wise 3D range image registration in forensic anthropology. Soft Computing 11(9), 819–828 (2007)
8. Ballerini, L., Cordón, O., Damas, S., Santamaría, J., Alemán, I., Botella, M.: Identification by computer-aided photographic supra-projection: a survey. Technical Report AFE 2007-04, European Centre for Soft Computing (2007)
9. Bowyer, K.W., Chang, K., Flynn, P.: A survey of approaches and challenges in 3D and multi-modal 3D + 2D face recognition. Computer Vision and Image Understanding 101, 1–15 (2006)
10. Bäck, T., Fogel, D.B., Michalewicz, Z. (eds.): Handbook of evolutionary computation. IOP Publishing Ltd and Oxford University Press (1997)
11. Laguna, M., Martí, R.: Scatter search: methodology and implementations in C. Kluwer Academic Publishers, Dordrecht (2003)
12. Santamaría, J., Cordón, O., Damas, S.: Evolutionary approaches for automatic 3D modeling of skulls in forensic identification. In: Giacobini, M. (ed.) EvoWorkshops 2007. LNCS, vol. 4448, pp. 415–422. Springer, Heidelberg (2007)
13. Calisti, M.: Ricostruzione 3D di teschi attraverso caratteristiche euristiche per applicazioni forensi. Master's thesis, University of Florence (2008)
14. Yoshizawa, S., Belyaev, A., Seidel, H.: Fast and robust detection of crest lines on meshes. In: 2005 ACM Symp. on Solid and Physical Modeling, pp. 227–232 (2005)

3D Processing and Visualization of Scanned Forensic Data

Alexander Ehlert[2] and Dirk Bartz[1,2]

[1] Visual Computing (ICCAS), University of Leipzig
dirk.bartz@medizin.uni-leipzig.de
[2] Visual Computing for Medicine Group, University of Tübingen

Abstract. A major task in forensic pathology is the documentation of surface injuries. In this contribution, we present a semi-automatic approach for the processing of data from 3D photogrammetry for the visualization of the body surface.

1 Introduction

3D photogrammetry is a relatively recent method to acquire three-dimensional data on the surface of a body to document lesions. Typically, a 3D surface scanner based on the structured light approach is positioned such that it captures the respective surface parts of a body. Due to the limited scanning volume of these scanners, several scans must be performed to capture a full body. This fact results in a labor-intensive process to combine the scans into a unified model of the body's surface. Furthermore, due to a scanning process inherent noise, the acquired data is far from being a perfect representation of the scanned body; sampling errors and color variations due to different lighting conditions are among the most typical problems. At the Institute of Forensic Medicine at the University of Tübingen, the scanned data was processed by a series of off-the-shelf software for image processing and rendering [25]. As indicated above, this process was very time-consuming and resulted in still insufficient quality.

Here, we describe a semi-automatic processing pipeline that employs various filtering techniques to identify and remedy sampling errors and to unify the color appearance due to variations of the lighting conditions. In particular, this pipeline consists of removing geometric data that does not belong to the scanned body (e.g., the post mortem table and vacuum mattress) and to correct lighting and color information. Finally, the various parts of the scans must be registered and combined into a unified model of the whole scanned body. In this process, the approach aims at maintaining the data fidelity and data integrity, which sometimes are competing goals.

Related Work

3D scanned data in forensic sciences has been used in a large variety of situations. In most cases, however, data from computed tomography (CT) and

S.N. Srihari and K. Franke (Eds.): IWCF 2008, LNCS 5158, pp. 70–83, 2008.

magnetic resonance imaging (MRI) has been used to scan a body and to derive forensic findings from it. Harris, for example, used 2D MRI images to provide evidence impact injuries [8], while Oliver et al. extended the approach to 3D reconstructions to represent the path of a bullet from CT data [14]. Thali et al. showed in a preliminary study of 40 forensic cases that post-mortem CT or MRI based forensic examinations were equal or in some aspects better than traditional autopsies [27]. Furthermore, he advocated for a full-body scan for a virtual autopsy. On the technical side, 2D slide images and 3D surface (SSD) visualizations of bone structures were used. For a more specific analysis of the sudden infant death syndrome, Preim et al. provided segmentation and analysis tools for volumetric data from CT/MRI [18], which enables a more refined analysis and visualization of the datasets. A different rendering approach was presented by Ljung et al. [12], who used high quality direct volume rendering to explore the post-mortem CT scans and to visualize the cause of death.

A related approach is post-mortem scanning of human bodies, like the Visible Human Project [13]. Next to CT and MRI scans, digitized anatomical cryosections are provided, which represent a high resolution, full color representation of the interior body parts.

Persson recently described different research areas for virtual autopsies, including the scanning of the body surface based on 3D photogrammetry-based scanning [16]. Brüschweiler et al. used such an approach to document lesions of local body surface areas [6] and Thali et al. combined this representation with the above described CT/MRI-based approach [26]. In contrast to this approach, we aim at a surface scanning approach to document the full body surface [7].

A very related research field are scanning projects in general, which are among the most researched topics in computer graphics and vision. Typical examples of such studies are the IBM Pietà project [9,22,4] and Stanford's Digital Michelangelo project [11], where numerous processing steps are applied [3]. Similar to our project, those scanned object do not fit into the scanning range of an individual scan and must hence be partially scanned. In this case, a new issues arise, since the 3D surfaces of the partial scans must registered [5] and the respective lighting and color information in the acquired textures must be corrected [1,21]. In particular the latter processing steps are required, since individual scans are subject to different lighting conditions, causing variations of intensity and color in the textures.[1] Once the textures are corrected, the polygonal model must be constructed [2,22], if not the alternative point-based rendering approach is used [23,17].

Lighting and color correction are already complex tasks that must balance correction quality and original information fidelity. Original information fidelity is important for all forensic documentation purposes. Unfortunately, partial scans will almost always be subject to different lighting conditions and hence have different color casts, which render this tasks as particularly difficult. In this spirit,

---

[1] Textures describe the color information captured by the cameras. Every pixel of the camera view plane is represented by one texture pixel (=texel). Together with the point sample geometry, the respective color can be assigned to the point sample.)

the approach of Paris et al. to compute a skin reflectance map taking into account lighting-independent parts [15] significantly changes the original data, and is hence not suited for forensic documentation.

A more suitable approach for color correction was proposed by Reinhard et al. [19], who perform a color space conversion from RGB to XYZ, followed by a principal component analysis (PCA), and a transformation into a perception-driven $l\alpha\beta$ color space [20]. This approach is very related to our simpler approach, which was derived concurrently (starting in 1999) and is based on a RGB color space PCA [7].

2   Methods

As mentioned in the last section, several steps must be performed to extract meaningful models from the scanned data. Figure 1 provides an overview of the individual processing steps, which will be explained in more detail in the following sections. In particular we differentiate two processing stages after scanning; processing of the individual partial scans (Section 2.2), and the matching steps that focus on correlating the different partial scans to the common model (Section 2.3).

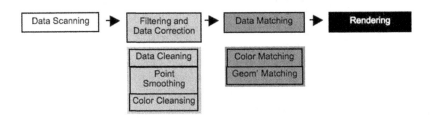

Fig. 1. Data processing pipeline

2.1   Scanning and Model Representation

The data acquisition is based on the coded or structured light approach [28], where a sequence of line patterns – which encode the x position in the image plane to simplify the correspondence problem – are projected onto the object to be scanned. Two CCD cameras in a stereoscopic setup capture the line patterns and acquire the surface information, which includes the geometry and the texture, through triangulation. The scanning, which resulted in the datasets used here, took place at the Institute of Forensic Medicine at the University of Tübingen, and at the Robert Bosch Hospital, Stuttgart, Germany.

The body is positioned on the post mortem table and a vacuum mattress is used to stabilize the body during scanning. This, however, also involves a repositioning of the body on the vacuum mattress to scan the lateral parts of the body, which in turn will cause different positions of the limbs, and typically of the body itself as well. Consequently, two different models for the front and the back part of the body will be constructed. To ensure a good coverage of the lesions, the scanner must be positioned such that each lesion is well-covered by the respective scan(s). This is particularly important, since parts of the body might be occluded by the vacuum mattress. Furthermore, sufficient overlap between neighboring scans must be ensured to provide sufficient common registration information. In difficult cases, marker objects may be attached to the body to improve the registration results.

The small scanning volume of the structured light scanner (with a view plane of approximately $1.0m \times 1.0m$) requires up to 28 to 40 individual scans to capture the full body, resulting in almost 400K sampling points per scan (each containing position, normal, and color), and hence in approximately 11M sampling points per model (front and back each). Due to the high number of acquired sample points, we chose a point-based representation, where each point consist of a position, a (texture-) color, and a normal. While a surface representation could be generated from this data, it would be a time-consuming and memory intensive operation. Furthermore, the intended documentation purpose would not really benefit from a surface representation, which involves additional interpolation and filter steps.

The resulting sampling points of the partial scans are sorted into an octree. Depending on the specified granularity, an octree block is split into child blocks, when the specified number of sample is exceeded. Empty child blocks (with no sample points) are removed. For better performance, the octree maintains only indices to the data points, which are stored in an array data-structure, and only the indices are involved into the construction process. After the construction process, the octree is traversed sequentially, and the sampling points of the traversed leaf blocks are sorted into a new array data-structure to maintain a better data locality that enables a more cache-sensitive accessing in the local neighborhood. The original array is then discarded.

Some of the processing steps of our reconstruction pipeline do require neighborhood information for the current data point. For that purpose, we use a k-nearest neighbor search in the octree, where each leaf block of the octree is sorted according to the distance to the current data point. Here, we can take advantage of the locality-sensitive data-structure, where the closest leaf blocks may be processed in the cache. Once the blocks are clearly outside of the k-nearest neighborhood, the respective subtree traversal is aborted. For all close-enough leaf blocks, the data points are stored in a distance map, which now enables the selection of the k closest elements that establish the k nearest neighborhood.

2.2   Data Correction

The first processing stage consists of three data correction steps that focus on the individual partial scans.

The large numbers of sample points from the 3D scans renders the handling of each individual point impractical from performance and user-interface points-of-view. Hence, we base our interaction on the leaf blocks of the octree instead of the individual points. The chosen granularity of the octree is a trade-off between

Fig. 2. Block editing of different partial scan of Accident Victim 1. All red blocks of the frontal (left) and lateral view (right) are part of the vacuum mattress and are selected for removal using region growing and cuboid tools.

sample point control (low granularity) and speed (high granularity). In practice, a block resolution of 4-5mm in each direction provided a good granularity for all purposes.

The first correction step – data cleaning – removes sample points that do not belong to the body, e.g. the vacuum mattress (see Fig. 2). In order to select the respective octree blocks, three selection operations are supported. **Picking** selects all blocks through the mouse taking into account the current screen position. **Frame selection** extends that concept to a whole screen area. Finally, the **region growing** operation selects all points that are located within an ϵ-proximity of already selected blocks. Since these operation sometimes select blocks representing body parts, an **undo** operation is also supported. In addition, the **cuboid tool** removes all blocks outside of a selected space region, and the **minimum filter** aims at outlier sample points by removing all leaf blocks that contain less than a specified number of data points.

The next step provides yet another mechanism to remove outliers and surface noise. Here we fit a tangent plane P to the local point neighborhood of every sample point and subsequently project the sample point onto that tangent plane using the Moving Least Square approach, generating a largely smooth surface [10]. In this smoothing process, we can at the same time reduce the number of sample points, where the fitness of a tangent plane is insufficient. Figure 3 shows an example of this procedure, where 80% of the sample points are maintained.

Due to noise in the measurement process (mostly camera noise), local colors need to be cleansed too. In order to maintain color fidelity, we only correct the lightness component in the HLS color space, and keep hue and saturation

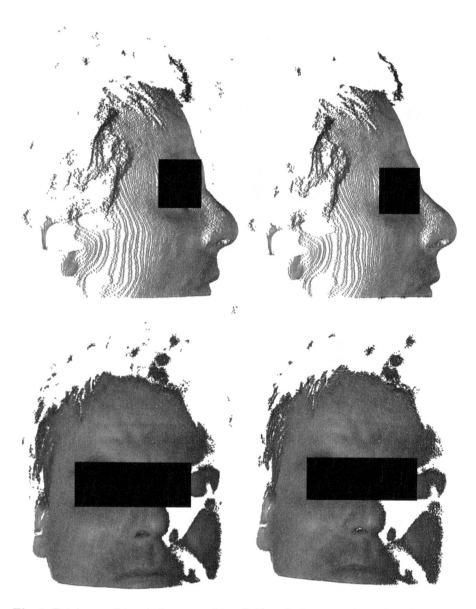

Fig. 3. Point smoothing: before smoothing (left) and after smoothing (right). Surface noise and outliers are removed, while data fidelity is maintained.

values. The lightness value is also the color component that is most affected by the camera noise. We use a median filter that adapts the lightness value of the sample points outside of the defined range. Alternatively, these points can also be removed. While the lightness-only modification limits possible color changes, this filtering step has to be applied with great care, since color fidelity are important for forensic applications. Figure 4 demonstrates the effect of color

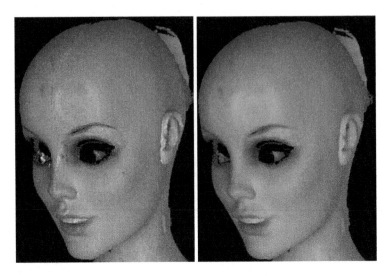

Fig. 4. Color cleansing of the Mannequin dataset - before (left) and after (right) modification of the lightness component in the HLS color space using a median filter

cleansing, where noise is removed. In particular the sample points near the right eye are successfully corrected.

2.3   Data Matching

So far, the individual partial scans have been cleansed from noise and outliers, and their geometry and color information has been corrected. The next processing stage focuses on color and geometry consistency between the partial scans. In that process, the color casts of the individual partial scan textures are adapted and the overlapping geometry of the point clouds of the partial scans are matched. In particular the first step involves a decision which of the partial scans will pose as the reference color cast for the other partial datasets.

In theory, all partial scans should exhibit the same color casts. This requires uniform lighting conditions, where a sufficient number of diffuse light sources are positioned around the body. Furthermore, a white fader should ensure that the different CCD cameras are calibrated to the same white to reduce the color capture differences. Unfortunately, scans are typically acquired in a less perfect environment, where the cameras are not white-calibrated and lighting exposes different conditions from scan to scan, leading to different color casts (Fig. 5) and diffuse brightness highlights (Fig. 6 left).

Some matching steps to correct the lighting differences can be quite simple; the varying brightness, for example, can be corrected by stretching the white and black points of the cameras in the histogram. This results in more even brightness ranges. Furthermore, gamma corrections provides for another brightness improvement. Note however, the partial scans often include different background

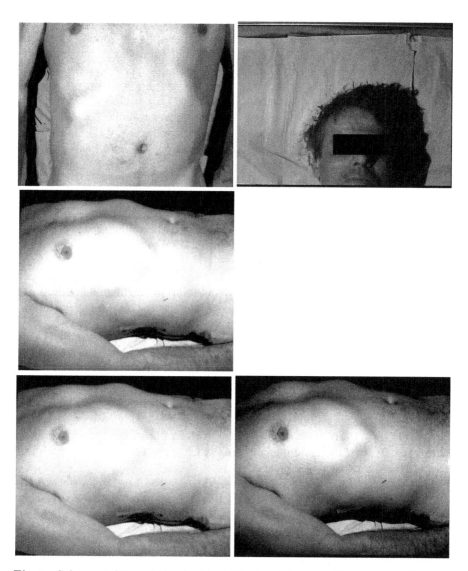

Fig. 5. Color matching of the Accident Victim 1 dataset: The top row shows two different reference images, the middle row shows the target image with the distorted color cast, the bottom row shows the resulting images with the matched colors. The left column shows a reference image that does not contain much distorting background material and the impact of the small visible patches of the white cover are negligible. The right column shows an example where the white cover significantly distorts the color casts.

materials. In some cases, a white cover will be visible (e.g., around the head), while it is not visible in other scans. This will change the histogram ranges of the texture information of the partial scans and must be taken into account.

The other necessary correct step involves the color itself. As mentioned above, different lighting conditions lead to different color casts. Figure 5 shows an example, where similar body regions expose different color tones, for example a more white or a more yellowish color. In our approach, we apply a principal component analysis (PCA) to the image to achieve a parameterization in a near-orthogonal color space. By transforming the color space of the target dataset (texture) to the color space of the reference dataset, we achieve a very similar color cast (see Fig. 5 bottom row). During this re-parameterization, all texels of the texture are considered. This, however, can lead to an ill-suited color casts in cases, where other materials (e.g., (white) covers, vacuum mattress, etc.) cover a significant area of the texture with a significantly different colors. A white cover, for example, will lead to a paler than usual color cast. In those cases, it is advisable to consider only areas of the texture that are covered by the body.[2] Note that this method is independent from the orientation of the different datasets – if for example one scan has a frontal/vertical orientation and the other scan has a lateral/horizontal orientation (Fig. 5).

So far, the textures still contain view point depending information, since diffuse lighting information – such as shadows and surface highlights – is still included due to the texture capturing process. In detecting those areas, we follow Rushmeier et al. [21], who examine all pixels with very low (shadows) or very high (highlights) pixel intensities. We then lookup the associated data points and estimate surface normals based on the local point neighborhood. Based on these normals, the approximate light position and its respective diffuse lighting effect for the data points can be computed and deducted from the texels intensity values. To smooth the appearance, the local texels are then corrected by interpolating the intensity values from its (possibly already corrected) neighborhood (Fig. 6 left and center left). The perceived flatness of the right image indicates the loss of directional information, which is then re-gained by mapping the texture to the respective point geometry.

While this color matching step is performed logically after data correction, it must in practice be done before it. The reason for this constraint is the correlation between point samples and texture coordinates, which is lost after some point samples are removed or transformed.

Once the textures of the partial scans are consistent, we need to match the pointset geometries of the partial scans. As we have mentioned before, the scans will be combined into two models; one for the front, one for the back, since the re-positioning of the body in the vacuum mattress would results in significant deformations, due to the movement of body parts (e.g., arms, legs). This way, however, we get away with a rigid registration approach. We chose a variation of the Iterative Closest Point algorithm (ICP) [5,24], where a random set of data points r_i is selected from the reference dataset and a matching pointset t_i is

---

[2] Theoretically, this can also happen with significantly color differences of the body, e.g., a large lesion. In those cases, which rarely have a significant impact, the irrelevant parts must be cropped from the texture. Pieces of clothes are rarely worn by the subjects at this point.

Fig. 6. Lighting correction with an image of the Accident Victim 1 dataset: Uncorrected image (left), corrected image (center right). Registration of two partial scans (yellow and blue) after good initial positioning. Before (center right) and after (right) ICP registration. The overlap demonstrates the improvement due to the registration.

searched in the target dataset. For an ideal match, we need to solve the linear equation system described in Equation 1, where t_i represent the perfect match of the respective data points in the target set.

$$M \cdot r_i = t_i \tag{1}$$

This, however, is not a realistic assumption, since measurement noise and our processing filter have changed the pointsets. Hence, we aim at minimizing the distance between reference and target pointsets in reference to the initial positioning. For this minimization step, we need to optimize six degrees of freedom; three degrees for a rotation, three degrees for a translation, realizing a rigid registration process. The optimization itself is computed by the Gauss-Newton algorithm, which minimizes distance D between reference points r_i and target surface $P(t_i)$, with $P(t_i)$ as an approximation of a tangential plane in an ϵ-neighborhood around t_i:

$$D = \sum_i dist(M \cdot r_i, P(t_i))^2 \tag{2}$$

The two right-most images of Figure 6 dhows how the initial position of two partial scans of the head is improved by the registration algorithm. The whole process is demonstrated in Figure 7 with the Accident Victim 1 dataset. Here, four partial scans of the head are registered and combined into one model.

As mentioned above, the minimized distance is based on the initial position of the target pointset. Hence, reference and target datasets need to be roughly pre-positioned, which we achieve by associating the partial scans with the respective parts of a dummy body model. If the structured light scanner itself is tracked, this information can be used to automate the pre-positioning process.

2.4 Rendering

Due to the large number of data points from the scanning, we opted against a polygonal surface rendering approach and chose instead a point rendering

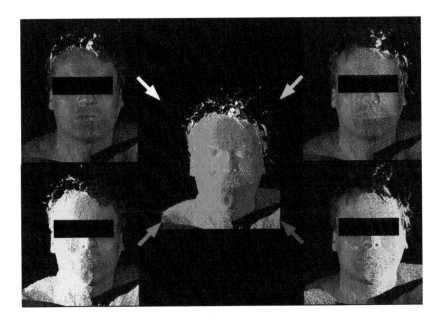

Fig. 7. Registration of partial scans: In this example, four partial scans of the frontal head are registered into a combined model. The different gray shades of the central image visualize the contributions of the different partial scans.

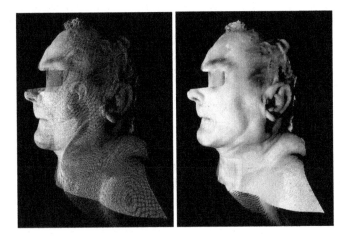

Fig. 8. Head region of accident victim 2 with small and large QSplats. The eye region has been removed for anonymity.

approach. Specifically, we adopted the QSplat approach of Rusinkiewicz and Levoy [23], which in essence use anti-aliased OpenGL points as rendering primitive for each data point of the final model. During the depth-sorted traversal of the octree, approximately eight million points are selected for rendering.

Standard OpenGL lighting and shading is used, whereas the points are weighted with their (corrected) color. Transparency blending along with variations of the OpenGL point sizes, and/or brightness is used to facilitate anti-aliasing. In cases where the respective point parameter OpenGL extension is not available, only the point size is varied. Figure 8 shows the effect of the varying point sizes from a very close point of view. Note that data points within one octree block are not depth sorted (only the blocks are depth sorted) to avoid excessive computational costs. The resulting visual defects, however, are negligible.

3   Results

The presented models of a Mannequin and of two accident victims, which consist each of 28 partial scans for dorsal and face-down body orientation. The latter models are the more challenging datasets, as they represent a real application situation, where scanning artifacts results in significant color and geometry deviations.

The individual processing steps applied to the partial and combined scans take approximately 1 minute, whereas the average total processing time of one subject takes about an hour, depending on individual tuning and manual cleaning requirements. This, however, is a significant less than the previous approach based on commercial off-the-shelf tools required [25].

Rendering of the full models took between 3 and 20fps depending on the size (number of point) of the models and the chosen resolution on a now outdated PC equipped with 3GB of main memory, an Intel P4@3GHz CPU, and an NVIDIA GeForce 6800 graphics accelerator.

4   Summary

In this contribution, we present our approach for the processing and rendering of partial scans into a combined overall model. A pipeline of processing steps has been described that includes a variety of cleaning and smoothing steps to remove outliers and noise, and a number of matching steps to combine the partial scans into one consistent model. Some of these processing steps are parameter-free and hence automatic, while other involve significant manual work.

The major goal of this approach was to provide an additional avenue for the documentation of surface injuries by fall, shock, or by a blow. It is not meant at replacing the currently used information, but to supplement the traditional sketches and photography. In the future, the acquired data is also intended for analysis tasks, e.g., the mechanism of accidents.

Future work will in particular focus on process workflow improvements, as most manual work can be reduced by it. This will in particular involve overview functionality to quickly associate partial scans with the respective body part. Furthermore, automatic pre-positioning can be achieved, if the scanner is successfully tracked in the environment.

Acknowledgments

All presented datasets are courtesy of the Institute of Forensic Medicine of the University of Tübingen. We thank Prof. Dr. Heinz-Dieter Wehner for the kind support of the project.

References

1. Bernardini, F., Martin, I., Rushmeier, H.: High-Quality Texture Reconstruction from Multiple Scans. IEEE Transactions on Visualization and Computer Graphics 7(4), 318–332 (2001)
2. Bernardini, F., Mittleman, J., Rushmeier, H., Silva, C., Taubin, G.: The Ball-Pivoting Algorithm for Surface Reconstruction. IEEE Transactions on Visualization and Computer Graphics 5(4), 349–359 (1999)
3. Bernardini, F., Rushmeier, H.: The 3D Model Acquisition Pipeline. Computer Graphics Forum. 21(2), 149–172 (2002)
4. Bernardini, F., Rushmeier, H., Martin, I., Mittleman, J., Taubin, G.: Building a Digital Model of Michelangelo's Florentine Pietà. IEEE Computer Graphics and Applications 22(1), 59–67 (2002)
5. Besl, P., McKay, N.: A Method for Registration of 3-D Shapes. IEEE Transactions on Pattern Analysis and Machine Intelligence 14(2), 239–256 (1992)
6. Brüschweiler, W., Braun, M., Dirnhofer, R., Thali, M.: Analysis of Patterned Injuries and Injury-Causing Instruments with Forensic 3D/CAD Supported Photogrammetry (FPHG): An Instruction Manual for the Documentation Process. Forensic Science International 132(2), 130–138 (2003)
7. Ehlert, A., Salah, Z., Bartz, D.: Data Reconstruction and Visualization Techniques for Forensic Pathology. In: Data Visualization (Proc. of Eurographics/IEEE VGTC Symposium on Visualization), pp. 323–330, 379 (2006)
8. Harris, S.: Postmortem Magnetic Resonance Images of the Injuried Brain: Effective Evidence in the Courtroom. Forensic Science International 50(2), 179–185 (1991)
9. IBM Research. Pietà Project (re-accessed, 2005), http://www.research.ibm.com/pieta/index.html
10. Levin, D.: Mesh-independent Surface Interpolation. In: Brunnett, G., Hamann, B., Müller, H., Linsen, L. (eds.) Geometric Modeling for Scientific Visualization, pp. 37–49. Springer, Heidelberg (2003)
11. Levoy, M., Pulli, K., Rusinkiewicz, S., Koller, D., Pereira, L., Ginzton, M., Anderson, S., Davis, J., Ginsberg, J., Curless, B., Shade, J., Fulk, D.: The Digital Michelangelo Project: 3D Scanning of Large Statues. In: Proc. of ACM SIGGRAPH, pp. 131–144 (2000)
12. Ljung, P., Winskog, C., Persson, A., Lindström, C., Ynnerman, A.: Full Body Virtual Autopsies using a State-of-the-art Volume RenderingPipeline. In: Proc. of IEEE Visualization, pp. 869–876 (2006)
13. National Library of Medicine. The Visible Human Project (re-accessed, 2007), http://www.nlm.nih.gov/research/visible/visible_human.html
14. Oliver, W., Chancellor, A., Soitys, J., Symon, J., Cullip, T., Rosenman, J., Hellman, R., Boxwala, A., Gormley, W.: Three-dimensional Reconstruction of a Bullet Path: Validation by Computed Radiography. Journal of Forensic Sciences 40(2), 321–324 (1995)

15. Paris, S., Sillion, F., Quan, L.: Lightweight Face Relighting. In: Proc. of Pacific Graphics, pp. 41–50 (2003)
16. Persson, A.: Virtual Autopsies Guide Postmortem Investigation. Diagnostic Imaging Europe (2), 20–28 (2007)
17. Pfister, H., van Baar, J., Zwicker, M., Gross, M.: Surfels: Surface Elements as Rendering Primitives. In: Proc. of ACM SIGGRAPH, pp. 335–342 (2000)
18. Preim, B., Cordes, J., Heinrichs, T., Jachau, K., Krause, D.: Quantitative Bildanalyse und Visualisierung für die Analyse von post-mortem Datensätzen. In: Proc. of Workshop Bildverarbeitung in der Medizin, pp. 6–10. Springer, Heidelberg (2005)
19. Reinhard, E., Ashikhmin, M., Gooch, B., Shirley, P.: Color Transfer Between Images. IEEE Computer Graphics and Applications 21(5), 34–41 (2001)
20. Ruderman, D., Cronin, T., Chiao, C.: Statistics of Cone Responses to Natural Images: Implications for Visual Coding. Journal of the Optical Society of America 15(8), 2036–2045 (1998)
21. Rushmeier, H., Bernardini, F.: Computing Consistent Normals and Colors from Photometric Data. In: Proc. of International Conference on 3D Digital Imaging and Modeling, pp. 99–108 (1999)
22. Rushmeier, H., Bernardini, F., Mittleman, J., Taubin, G.: Acquiring Input for Rendering at Appropriate Levels Of Detail: Digitizing a Pietà. In: Proc. of Eurographics Symposium on Rendering, pp. 81–92 (1998)
23. Rusinkiewicz, S., Levoy, M.: QSplat: A Multiresolution Point Rendering System for Large Meshes. In: Proc. of ACM SIGGRAPH, pp. 343–352 (2000)
24. Rusinkiewicz, S., Levoy, M.: Efficient Variants of the ICP Algorithm. In: Proc. of International Conference on 3D Digital Imaging and Modeling, pp. 145–152 (2001)
25. Supke, J., Wehner, H., Szczepaniak, S.: Streifenlichttopometrie (SLT): A New Method for the Three-dimensional Photorealistic Forensic Documentation in Colour. Forensic Science International 113(1-3), 289–295 (2000)
26. Thali, M., Braun, M., Wirth, J., Vock, P., Dirnhofer, R.: 3D Surface and Body Documentation in Forensic Medicine: 3-D/CAD Photogrammetry Merged with 3D Radiological Scanning. Journal of Forensic Sciences 48(6), 1356–1365 (2003)
27. Thali, M., Yen, K., Schweitzer, W., Vock, P., Boesch, C., Ozdoba, C., Schroth, G., Ith, M., Sonnenschein, M., Doernhoefer, T., Scheurer, E., Plattner, T., Dirnhofer, R.: Virtopsy, a New Imaging Horizon in Forensic Pathology: Virtual Autopsy by Postmortem Multislice Computed Tomography (MSCT) and Magnetic Resonance Imaging (MRI) - a Feasibility Study. Journal of Forensic Sciences 48(2), 386–403 (2002)
28. Wahl, F.: A Coded Light Approach for 3-Dimensional (3D) Vision. Technical Report IBM RZ 1452 (52546), IBM Zurich Research Laboratory (1984)

Person Identification Based on Barefoot 3D Sole Shape

Michael Hild

Osaka Electro-Communication University, Faculty of Information and
Communication Engineering
Osaka, Neyagawa 572-8530, Hatsu-cho 18-8
hild@hilab.osakac.ac.jp

Abstract. In this paper we report on person identification based on the three-dimensional shape of the human foot sole. After acquisition of 3D foot sole data with a laser scanner, data normalization is performed and person identification is carried out using principal component analysis (PCA) and linear discriminant analysis (LDA) in conjunction with 1–NN classification, and support vector machines (SVM) in conjunction with output value thresholding. The obtained high identification and rejection rates indicate that the 3D foot sole shape of the human bare foot is adequate for person identification.

1   Introduction

In recent years, person identification methods based on the measurements of human body parts such as the finger tips (finger print pattern), the human face (facial intensity pattern), the human iris (iris pattern), the human hand palm (palm print pattern), hand veins (vein pattern), and the human ear (ear shape pattern) have been explored extensively and various identification techniques have been proposed by the biometric research community [1,2,3,4,5]. In addition to these methods, in the field of forensics, blood type testing has been used primarily for ruling out identities of a person, and quite recently DNA testing has been used for determining person identity under difficult circumstances. The human *foot sole* as a possible body part to be used for person identification has also received some attention. However, to date these efforts have focused on the *2D footprint's* global region shape and the wrinkles and epidermal ridges on the foot sole surface, which closely resemble the characteristic surface pattern that exists on the hand palms (which have been used in legal investigations for positive identification of individuals). [6,7,8]

In this paper we report on our work on person identification based on the *three-dimensional shape of the human foot sole*. After acquisition of 3D foot sole data with a laser scanner, we perform data normalization and person identification using three different classification methods: principal component analysis (PCA) followed by nearest neighbor classification, linear discriminant analysis (LDA) followed by nearest neighbor classification, and support vector machines

S.N. Srihari and K. Franke (Eds.): IWCF 2008, LNCS 5158, pp. 84–95, 2008.
© Springer-Verlag Berlin Heidelberg 2008

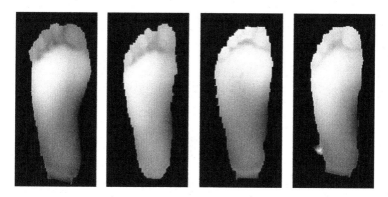

Fig. 1. Examples of foot sole range images

(SVM) followed by output thresholding. The purpose of this research is to determine whether the human foot sole is adequate for person identification, and to determine which one of the three classification methods would give the best identification results with this kind of data.

This paper is organized as follows: In section 2, the foot sole data acquisition method is described, in section 3, the three classification methods used for foot sole identification are explained. In section 4, experimental results are presented, and conclusions are drawn in section 5.

2   Foot Sole Data Acquisition

2.1   Data Acquisition Method

The foot sole data is acquired using a laser scanner which takes three-dimensional measurements $\vec{M}_i = (M_i \quad Y_i \quad Z_i)^T$ of scene surface points. The measurements are obtained as a lattice of equi–angular intervals in vertical and horizontal directions.

For scanning the foot sole surface, we place the subject's leg on a horizontal plane in front of the laser scanner, with the toes pointing upwards; then the foot sole surface is adjusted to vertical orientation by having it rest against a vertical metal plate, which is temporarily placed on the ground plane between foot sole and laser scanner. In this way the foot sole becomes perpendicular to the laser scanner's Z-direction. The vertical metal plate is subsequently removed and the foot sole is scanned. A diagram of this scan setup is shown in Fig.2.

2.2   Computation of Foot Sole Range Image

In order to conveniently represent the 3D foot sole data for further processing, a *range image* is computed from each raw scan data set. The range image represents the 3D data as a 2D data array $D(x_i, y_i), \quad i = (1, 2, ...N)$, where the x_i- and y_i-variables correspond to the X- and Y-axes of the 3D scan coordinate

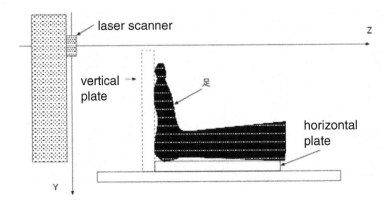

Fig. 2. Laser scan setup

system, respectively, and the Z-coordinate values are stored in $D(x_i, y_i)$. First, the 3D coordinate values X_i and Y_i are rounded to integer values x_i and y_i, respectively. Then the Z_i coordinate value is transformed such that it falls into value range $[0,255]$ using Eq.(1), and stored in array $D(x_i, y_i)$:

$$D(x_i, y_i) = 255 - [round(Z_i) - Z_{min}] \cdot (255/50) \qquad (1)$$

where Z_i satisfies $(Z_{min}) < Z_i < (Z_{min} + 50)$. Z_{min} is the Z-coordinate of the data point closest to the laser scanner. Only those foot sole surface points that occur within a $50mm$ interval beginning at Z_{min} are included in the range image as shown in Fig.3. The Z-coordinate values in the range image are measures of the deepness of the corresponding points on the image plane.

In order to fill the gaps between foot sole pixels in the range image occurring at this stage of processing, linear range value interpolation is carried out. First, range value linear interpolation in the direction of the x-axis is carried out between pixels where gaps are less than 3 pixels wide. Next, linear interpolation is carried out in the y-axis direction in order to fill gaps of less than 6 pixels width. Finally, the same processing is carried out again in x-axis direction, but now for gaps of less than 8 pixels width. In this way the interpolation remains simple and yet produces smooth surfaces.

The laser scans usually include part of the horizontal plane on which the foot rests. These pixels are removed by setting them to value 0 in the range image.

Since the foot sole range images are used as input data to classifiers, the orientations of all data sets have to be normalized with respect to rotation around some axis parallel to the Z-axis of the scanner coordinate system. First, the regions' centers of gravity and the angles between their principal axes and the X-axes are computed, and then the regions are rotated around the center of gravity so that the principle axes become vertical. Moreover, the regions are translated such that the centers of gravity coincide with the centers of the range images.

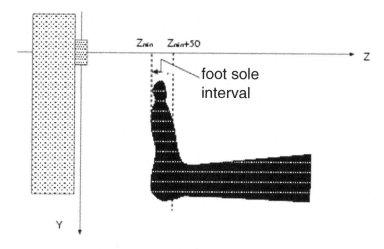

Fig. 3. Transformation depth interval of range image

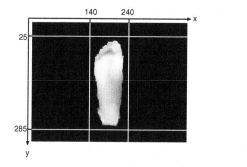

Fig. 4. Clipping of foot sole range image

Fig. 5. Example of normalized foot sole range image

Since rotation is a real–valued number operation performed on an integer array, pixel gaps due to value rounding will appear. In order to fill these gaps, range value interpolation is carried out once more.

Finally, for each image a rectangle slightly larger than the foot sole region is fixed with respect to the foot sole region center, and the thus defined rectangular region is clipped from the range image as is shown in Fig.4 and Fig.5. The size of the rectangle is kept constant for every acquired range image.

3   Classification of 3D Foot Sole Data

3.1   Classification with SVM

In this section, the support vector machine (SVM)-based classification method for person identification based on 3D foot sole data is described. Since the SVM

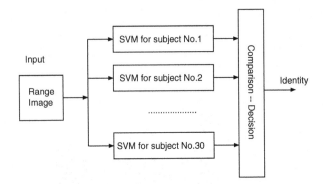

Fig. 6. Structure of SVM-based person identification system

is a binary classification method, one SVM classifier is trained for each person registered in the database, which results in one file of *model data* per person containing the person's support vectors.

In identification mode, when the person identity for a given *test range image* is to be determined, the *test range image* is fed into each one of the trained SVMs, and the SVMs' output values are evaluated in order to reach the identification decision. A diagram of this system is shown in Fig.6. A *positive* output value from a given SVM is interpreted as indicating that the input data had been scanned from the foot of the person for which this particular SVM had been trained. On the other hand, a negative output value indicates that the scan data is from some different person. For output values close to zero, the result is interpreted as indeterminate. When multiple SVMs produce positive output values, the SVM with the highest value is selected as the classifier indicative of person identity.

Each SVM classifier in this system is trained as follows: Assuming that the SVM is to be trained for a given person A, the same amount of foot sole scan data from person A and scan data from all other persons registered in the database are included in the training data set. In this case, the data from person A are labeled as *positive* data, whereas all other data are labeled as *negative* data.

3.2   Classification with PCA and 1–NN

Principle components analysis (PCA) is one of the many multivariate analysis techniques; it reduces high–dimensional input data to low–dimensional data before classification of the data takes place. [9] It filters out information that has little relevance to the classification task and thereby compresses the input data set.

All range images of size $W \times H$ are converted to N-dimensional 1D data vectors $\boldsymbol{D}_{ij}$, where $N = W \times H$ and subscripts (ij) denote the j-th image of the i-th person. The mean vector $\boldsymbol{F}$ is computed over all range images used for training:

$$\boldsymbol{F} = \frac{1}{M} \sum_{i,j}^{M} \boldsymbol{D}_{ij} \tag{2}$$

The mean vector is subtracted from each data vector, $G_{ij} = D_{ij} - F$, and the covariance matrix C of the difference vectors G_{ij} is computed:

$$C = \frac{1}{M} \sum_{i,j}^{M} G_{ij} \cdot G_{ij}^{T} \tag{3}$$

In order to acquire the principle components of the input data, the eigenvalues E_i and the corresponding eigenvectors V_i of the covariance matrix are computed. The N eigenvalues $E_i, (i = 1, 2, \ldots, N)$ sorted in descending order are denoted as $E_i', (i = 1, 2, \ldots, N)$ and the first $N_0 < N$ eigenvalues $E_j', (j = 1, 2, \ldots, N_0)$ and their corresponding eigenvectors V_j are selected as principle components.

Number N_0 is determined such that a number L just exceeds value 0.95 as more components E_j' are added. L is computed as

$$L = \frac{\sum_{j=1}^{N_0} E_j'}{\sum_{j=1}^{N} E_j'} \tag{4}$$

The selected N_0 eigenvectors $V_i = (v_{i1} \ldots v_{iN})^T, (j = 1, 2, \ldots, N_0)$ are used as the base vectors of a N_0-dimensional vector space, which is used to reduce the originally N-dimensional data D_{ij} to N_0-dimensional data vectors D_{ij}' by projecting the original data vectors onto the low–dimensional vector space using Eq.(5):

$$D_{ijk}' = D_{ij} \cdot V_k, (k = 1, 2, \ldots, N_0) \tag{5}$$

where D_{ijk}' denotes the k-th component of D_{ij}'.

In order to determine which class a given input data vector belongs to, we carry out 1–NN (nearest neighbor) classification in the N_0–dimensional eigenvector subspace. In order to be able to directly compare PCA–based foot sole classification results with SVM–based foot sole classification results, we design the PCA-based nearest neighbor classification process as a binary classification scheme similar to the SVM classification scheme (see section 3.1). The same data sets as were used for SVM–classifier training are used for PCA 1–NN classification, too, and one binary 1–NN classifier per registered person is trained. One separate eigenvector subspace for each registered person is computed; this in general leads to subspaces that differ in their dimensions N_0.

In the i-th person's subspace, the training data is labeled with respect to their belonging to either one of the two possible classes *data of the i-th person* or *data of all other registered persons*. When an arbitrary, not yet identified data vector is tested as to whether it belongs to the i-th person class or not, the distances Q_{ij} between this data vector and all training data vectors in both constituent classes is calculated, and the class of the vector with the smallest distance is taken as indicating the resulting class.

3.3 Classification with LDA and 1–NN

The third method for foot sole classification used in this study, Linear Discriminant Analysis (LDA) [9] in conjunction with a 1–NN classifier, configures the subspace such that the classes can be optimally linearly separated, with the separating hyperplane given by Eq.(6):

$$z = a_0 + a_1 D_{ij1} + a_2 D_{ij2} + \ldots + a_N D_{ijN} = \boldsymbol{a}^T \cdot \boldsymbol{D}_{ij} \qquad (6)$$

$\boldsymbol{D}_{ij}$ is the i-th person's j-th range image data vector, $\boldsymbol{a}$ is the linear discriminant vector that is to be determined, and N is the dimension of the data vectors. Denoting the training data's *between-classes covariance matrix* as C_B and the *within-class covariance matrix* as C_W, vector $\boldsymbol{a}$ is first determined by maximizing the correlation ratio

$$H = \frac{\boldsymbol{a}^T \cdot C_B \cdot \boldsymbol{a}}{\boldsymbol{a}^T \cdot C_W \cdot \boldsymbol{a}} \qquad (7)$$

with respect to vector $\boldsymbol{a}$. This step constitutes the training step of the method.

In identification mode, test data vectors $\boldsymbol{D}_{ij}$ are projected onto the linear discriminant vector $\boldsymbol{a}$, which results in the (scalar) discriminant scores z of Eq.(6). The discriminant scores z are optimally class–separated. In this 1D discriminant score space 1–NN classification is used to determine the class identity of the test data vector.

4   Experimental Results

Foot sole scans were obtained with a VIVID700 laser scanner (Minolta). It acquired the 3D coordinates of measurement points on human foot sole surfaces as a 200×200 points array. We scanned the foot soles of a total of 30 subjects in their early twenties, with 30 scans per subject. These data sets were then transformed into range images and normalized as described in section 2. The resulting foot sole range images were of size 25×49 pixels, which is equivalent to 1176-dimensional data vectors. These 900 range images were used for classifier training and identification experiments with all three classification methods.

4.1   Identification Based on SVM

For the SVM-based identification experiment we used the SVM^{light} [10] program in linear classification mode. Of the 30 range images that were scanned per person, 20 range images were used for training the SVM classifiers, and the remaining 10 images were used for the identification experiment, giving a total of 300 range images to be identified. The 20 training images per person were selected at random. A total of 30 SVMs were trained, one per person as described in section 3.1.

In order to test SVM identification performance of each trained SVM, we fed all 300 test images consecutively into each SVM and recorded the SVM output values. The foot sole identification results of the SVM of Subject No.1 is shown in Fig.7 as a graph "SVM output value versus test image number". Range images No.1 - 10 from Subject No.1 and the positive SVM responses to these input images indicated "correct identification". Test images No.11 - 300 were

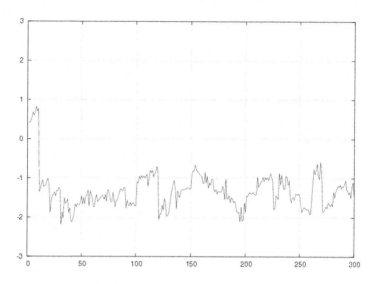

Fig. 7. Output values for SVM of Subject No.1

the images for which this SVM was not trained, and negative SVM responses indicated "correct rejection". In this case, 100% correct identification of Subject No.1's images as well as 100% correct rejection of images from all other subjects was achieved. As another example, the graph of Subject No.2's SVM is shown in Fig.8; it is similar, only that in this case Subject No.2's range images occupy the No.11 - 20 interval on the horizontal axis (where one would expect positive output values). The result in this case, too, was 100% correct identification as well as 100% correct rejection. For all the other subjects we obtained similar results of perfect identification and rejection.

Since 100% identification and rejection rates at the present state of the art are hardly ever achieved with other types biometric data, we wanted to test how robust these results are against moderate modifications of the data sets. For this purpose we prepared the five test range images shown in Fig.9, which include the cases of *holes in foot sole region, lower half of foot sole region only, upper half of foot sole region only, missing parts of foot sole region rim,* and *foot sole region border only.* We found that person identification is unaffected in the cases of *holes in foot sole region, upper half of foot sole region only,* and *foot sole region border only.* Person identification failed in the two cases of *lower half of foot sole region only* and *missing parts of foot sole region rim.* This leads to the conclusion that a significant portion of identification–relevant information is constituent in the foot sole region rim, and in particular in the upper region part.

As we were also interested in the long–term stability of the foot sole used as a biometric feature, after a 3–months intermission we re–scanned the foot soles of ten of the earlier subjects and repeated the identification experiment using five foot sole range images per subject. The result was 90% correct identification and 98.9% correct rejection, which is lower than the perfect rates obtained earlier. We

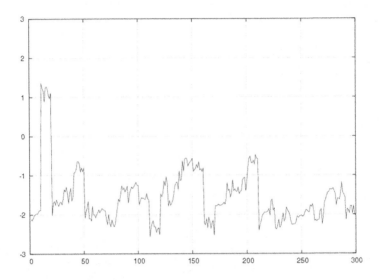

Fig. 8. Output values for SVM of Subject No.2

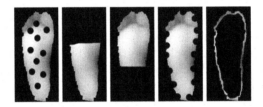

Fig. 9. Example of foot sole range images with five types of (artificially introduced) defects

conclude that the long–term stability of 3D foot sole–based person identification needs to be investigated further.

4.2   Identification Based on PCA and 1–NN

For the PCA–based identification experiment we used two different implementations of the PCA method: 1. the Jacobi method [13], and 2. the PCA implementation included in the R statistics package [11,12]. The test data sets were the same as those used for the SVM–based experiment, using 10 images per subject for the identification experiment, giving a total of 300 range images to be identified. A total of 30 PCA subspaces were computed from the training data, one subspace per person, as described in section 3.2.

In the identification phase, test data vectors were transformed into each of the 30 prepared subspaces. When a data vector from the i-th subject in the

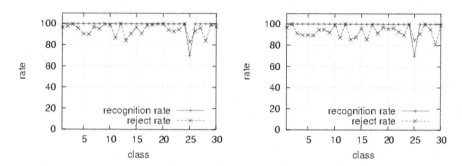

Fig. 10. PCA–based identification results (Jacobi method)
Fig. 11. PCA–based identification results (R statistics package)

i-th subject's subspace had as its closest neighboring training vector one of the i-th subject training vectors, then the identification was recorded as *correct*, otherwise incorrect. The number of correctly identified data vectors versus the total number of the i-th subject's data vectors (i.e. 10) gave the *recognition rate*. Likewise, when a data vector not belonging to the i-th subject in the i-th subject's subspace had as its closest neighboring training vector one of the training vectors *not* belonging to the i-th subject, the *rejection* was recorded as *correct*, otherwise incorrect. The *rejection rate* is defined as the number of correctly rejected data vectors versus the total number of data vectors (i.e. 290). The foot sole identification results obtained this way are shown in Fig.10 and Fig.11 as graphs labeled "Recognition rate versus class (i.e. range image number)" and "Rejection rate versus class". As can be observed, only subject No. 25 had three mis–identifications with both PCA implementations. The overall identification rate was 99.0%, and the overall rejection rates were 94.6% and 93.0% for the Jacobi method and R statistics package implementations, respectively.

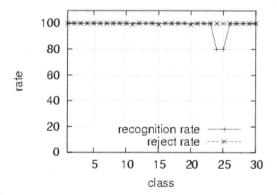

Fig. 12. LDA–based identification results

4.3   Identification Based on LDA and 1–NN

The implementation of the LDA algorithm used for the experiments was the LDA program included in the R statistics package [11,12]. The data sets were the same as those used for SVM– and PCA–based foot sole identification experiments. The foot sole identification results are shown in Fig.12 as a graph labeled "Recognition rate versus class (i.e. range image number)" and "Rejection rate versus class". As can be observed, only subjects No.24 and 25 had two mis-identifications. The rejection results turned out to be perfect.

The overall result for all 30 subjects and all 300 test images was a recognition rate of 98.7% and a rejection rate of 99.9%.

5   Conclusion

In this paper we reported on our work on person identification based on the three-dimensional shape of the human foot sole. After acquisition of 3D foot sole data with a laser scanner, we performed data normalization and person identification using three different classification methods: principal component analysis (PCA) and 1–NN, linear discriminant analysis (LDA) and 1–NN, and support vector machines (SVM) and output thresholding. All three methods produced recognition rates of about 99% and rejection rates of above 93%. This good identification performance of barefoot 3D foot sole shape came as a surprise to us as we had never achieved such high rates with 2D human face images using the same classification methods. Of course, our results must be viewed as only being preliminary, since we had included only 30 subjects and 300 range images in these experiments. Nonetheless, these results turned out to be exceptionally consistent across the three different classification methods used. In this sense we interpret these results as evidence that the 3D human foot sole is adequate for person identification. SVM classification gave slightly better identification results than the other classification methods.

Acknowledgements. I would like to express my appreciation of the following students who carried out the scanning of human foot soles and participated in the processing of many data sets: Keita Tai, Hideyuki Yamaji and Kazuhisa Matsuo. I would also like to express my gratitude for having received two very fair, competent and stimulating review reports. Unfortunately, I was not able to provide answers to all these suggestions in this paper due to time constraints.

References

1. Jain, A.K., Pankanti, S., Prabhakar, S., Hong, L., Ross, A.: Biometrics: A Grand Challenge. In: 17th International Conference on Pattern Recognition (ICPR 2004), vol. 2, pp. 935–942 (2004)
2. Zhao, W., Chellappa, R., Rosenfeld, A., Phillips, P.J.: Face Recognition: A Literature Survey. ACM Computing Surveys, 399–458 (2003)

3. Daugman, J.: Recognizing Persons by Their Iris Pattern. In: Jain, A.K., Bolle, R., Pankanti, S. (eds.) Biometrics: Personal Identification in Networked Society, pp. 103–121. Springer, US (1999)
4. Wu, X., Wang, K., Zhang, D.: Palmprint Authentication Based on Orientation Code Matching. In: Kanade, T., Jain, A., Ratha, N.K. (eds.) AVBPA 2005. LNCS, vol. 3546, pp. 555–562. Springer, Heidelberg (2005)
5. Chen, H., Bhanu, B., Wang, R.: Performance Evaluation and Prediction for 3D Ear Recognition. In: Kanade, T., Jain, A., Ratha, N.K. (eds.) AVBPA 2005. LNCS, vol. 3546, pp. 748–757. Springer, Heidelberg (2005)
6. Kennedy, R.B.: Barefoot Comparison and Identification Research, Technical Memorandum TM-12-95, Canadian Police Research Centre (1995)
7. Kennedy, R.B.: Uniqueness of bare feet and its use as a possible means of identification. Forensic Science International 82(1), 81–87 (1996)
8. Tong, L., Li, L., Ping, X.: Shape Analysis for Planar Barefoot Impression. In: Huang, D.-S., Li, K., Irwin, G.W. (eds.) ICIC 2006. LNCIS, vol. 345, pp. 1075–1080. Springer, Heidelberg (2006)
9. Bishop, C.M.: Pattern Recognition and Machine Learning. Springer, Heidelberg (2006)
10. Joachims, T.: Making large-Scale SVM Learning Practical. In: Schölkopf, B., Burges, C., Smola, A. (eds.) Advances in Kernel Methods - Support Vector Learning. MIT-Press, Cambridge (1999)
11. Venables, W.N., Ripley, B.D.: Modern Applied Statistics with S-PLUS, 3rd edn. Springer, Heidelberg (1999)
12. http://www.r-project.org/
13. Press, W.H., Vetterling, W.T., Teukolsky, S.A., Flannery, B.P.: Numerical Recipes in C, 2nd edn. Cambridge University Press, Cambridge (1992)

Computerized Matching of Shoeprints Based on Sole Pattern

Rui Xiao and Pengfei Shi

Institute of Image Processing and Pattern Recognition, Shanghai Jiao Tong University
200240 Shanghai, China
{rshell, pfshi}@stju.edu.cn

Abstract. Shoeprints are common clues left at crime scenes that provide valuable evidence in detecting criminals. Traditional shoeprints matching algorithm is based on manual coding with limited recognition ability, and the results are strongly dependent on the operator. In this paper, a shoeprint matching method based on PSD (power spectral density) and Zernike moment have been investigated. The PSD method aims at pressing images and the legible shoeprints. The correlation coefficients of the PSD value of each image are used as the measurements of similarity. In addition, the Zernike method has been developed for blurred crime scene shoeprint images and shoeprints with complex backgrounds. A series of irregular shapes are employed to identify the shoeprints. Features are then selected according to the Zernike moments of these shapes. More than 400 real shoeprint images have been tested, experimental results support that the method is effective in shoeprint matching.

Keywords: forensic science; shoeprint; Zernike moments; power spectral density; invariance.

1 Introduction

Shoeprints are the most common traces left at crime scenes that provide valuable evidence in the identification of criminals and in linking related cases. Traditionally, the classification of collected shoeprints is conducted manually by police officers and forensic scientists. It is a time consuming task due to the large amount of images in the database. Moreover it is hard for several operators to agree on a certain classification. So the development of a computer-based matching algorithm is extremely necessary to help investigators in rapid matching of the shoeprints and enhance the recognition rate, and thereby results in the sufficient use of shoeprint evidence.

A number of computer-based methods have already been proposed [1-5,11]. In [1], [2], shoeprints are described by a set of basic shapes, such as circles, rectangles, triangles, etc, by a forensic expert and these shapes are then used in the comparison of shoeprints. The problems of this method are that the recognition result is strongly dependent on the operator and that modern shoes has increasingly more complex patterns that are difficult to describe using a few basic shapes. In [11], fractals are utilizes to represent the shoeprints. They verified the translation invariance of the method, but rotation and scale invariance have not been reported in their work.

S.N. Srihari and K. Franke (Eds.): IWCF 2008, LNCS 5158, pp. 96–104, 2008.
© Springer-Verlag Berlin Heidelberg 2008

In this paper, a method based on PSD has been used to match press-down shoe-prints and high quality spot shoeprints [4]. First, the PSD value of each shoeprint image is calculated, and then the correlation coefficients are employed to evaluate the similarity between shoeprints. Since PSD has the property of translation and rotation invariance, the matching will not suffer from the difficult image registration process. A problem with this method is that its effectiveness is dependent on the quality of shoeprint images. When the image contains serious background noise or is hardly separated from shoeprint patterns, the results are always unsatisfactory.

For blurred crime scene shoeprint images and shoeprints with disordered back-grounds, we propose an algorithm based on Zernike moments. By extracting the shapes from shoeprints, the shoeprint matching is transferred into the matching of the extracted shapes. A good shape descriptor should have enough discriminating power and be invariant with respect to translation, rotation and scale of the shape. Invariant moment is one of the most extensively used shape descriptors since it was firstly introduced in the 60's by Hu [6]. Using combinations of geometric moments, Hu derives a set of 7 invariant moments to describe a shape. However, since the basis is not orthogonal, these moments suffers from a high degree of information redundancy. Because of these disadvantages, Teague [7] introduced orthogonal polynomials-based moments, Zernike moments. Teh [8] compared the noise sensitivity, information redundancy and representation capability of each proposed invariant moments and concluded that Zernike moment has the best performance. For this reason, we use Zernike moments as shape descriptor in this paper. Because the Zernike moments can be calculated irrespective of shape complexity, the method can be applied to the matching of shoeprints with any irregular patterns.

2   Shoeprint Matching Based on PSD

2.1   The Property of PSD

Let $f_1(x, y)$ presents a digital image whose Fourier transform is $F_1(u, v)$, and then its PSD $P_1(u, v)$ is defined as:

$$P_1(u, v) = |F_1(u, v)|^2 \tag{1}$$

The transformed image of $f_1(x, y)$ with translation (x_0, y_0) and rotation θ defines $f_2(x, y)$ as described in formula (2)

$$f_2(x, y) = f_1(x \cos \theta + y \sin \theta - x_0, -x \sin \theta + y \cos \theta - y_0) \tag{2}$$

The corresponding FT, $F_2(u, v)$ will be replaced by

$$F_2(u, v) = e^{-j2\pi(ux_0 + vy_0)} F_1(u \cos \theta + v \sin \theta, -u \sin \theta + v \cos \theta) \tag{3}$$

Then the PSD value of $f_2(x, y)$ can be described as:

$$P_2(u, v) = |F_2(u, v)|^2 = P_1(u \cos \theta + v \sin \theta, -u \sin \theta + v \cos \theta) \tag{4}$$

From formula (4), we can see that PSD has the translation invariance and the rotation of $f_2(x, y)$ results in the same rotation of the PSD.

2.2 Shoeprint Matching Based on PSD

Pre-procession like image enhancement and down-sampling is needed to reduce the computational load and achieve better result. A zero mean image is then obtained by subtracting the average gray level of the whole image from each pixel. The PSD value is then calculated according to formula (1) and their correlation coefficient is then considered to measure the similarity between two shoeprints [4].

$$\hat{P}_1(u,v) = [P_1(u,v) - mean(P_1)] / std(P_1) \tag{5}$$

$$\hat{P}_2(u,v) = [P_2(u,v) - mean(P_2)] / std(P_2) \tag{6}$$

$$r = \sum_i \sum_j \hat{P}_1(u,v)\hat{P}_2(u,v) \tag{7}$$

where $mean(f)$ stands for the average value of pixel values of f and $std(f)$ stands for the standard deviation.

3 Shoeprint Matching Based on Zernike Moments

3.1 Zernike Moments

In polar system, the n-order Zernike moment of an image is defined as:

$$Z_{nm} = \frac{n+1}{\pi} \int_0^{2\pi} \int_0^1 V_{nm}^*(r,\theta) f(r,\theta) r dr d\theta \tag{8}$$

where $V_{nm}(r,\theta)$ is the (n, m) order of the Zernike basis function defined over the unit disk:

$$V_{nm}(r,\theta) = R_{nm}(r)e^{im\theta} \tag{9}$$

And V_{nm}^* is a complex conjugate of V_{nm} .

The Zernike radial polynomials $R_{nm}(r)$ are defined as:

$$R_{nm}(r) = \sum_{s=0}^{(n-|m|)/2} (-1)^s \frac{(n-s)!}{s!(\frac{n-2s+|m|}{2})(\frac{n-2s-|m|}{2})!} r^{n-2s} \tag{10}$$

where n is a non-negative integer, and m is a nonzero integer subject to the following constraints: $n-|m|$ is even and $|m| \le n$.

To obtain scale and translation invariance, the image is first subjected to a normalization process. Set the origin of the coordinate system to the centroid of the shape, and then enclose the shape completely from the centroid to the outermost pixels of the shape by a square. In succession, resample the square to a pre-determined size by interpolation to reduce the computational load.

The rotation invariance of Zernike moments is straightforward to obtain. Suppose the rotation of an image is presented by an angle ϕ, the transformed Zernike moment function Z_{nm}^R can be expressed by:

$$Z_{nm}^R = Z_{nm} \exp(-jm\phi) \tag{11}$$

where Z_{nm} is the Zernike moment of the original image. Then the magnitude of Zernike moments has rotational invariance showed below:

$$Z_{nm}^R = | Z_{nm} \exp(-jm\phi) | = | Z_{nm} | \tag{12}$$

3.2 Feature Extraction

Extract the visible patterns on each shoeprint as shown in Fig. 1.

Fig. 1. Example of shoeprint impression and shape extraction

A shape normalization process is needed before calculating the Zernike moments. The shape is enclosed completely from the centroide to the outermost pixels by a square, and is then resampled to a pre-determined size by double-linear interpolation. This normalization method ensures both translation and scale invariance and preserves the ratio of the shape. Fig. 2 illustrates the normalization process and the establishment of the polar coordinate system.

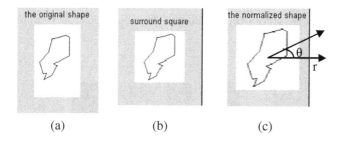

Fig. 2. Shape normalization and coordinate system

Through formula 8-10, Zernike moments of each shape can be easily calculated. Obviously, the higher order we extract, the more accurate we can represent a shape, but accordingly increase the computational expense and therefore slow down the operating speed. The orthogonality of Zernike moments enables the separated contribution of each order moment which makes the reconstructed image can be simply achieved by adding these individual contributions. What's more, owing to this characteristic of Zernike moments, the representation ability of each order moments can be evaluated through the comparison of the reconstructed image and the original image. This information is then used in the feature selection process [9].

Let $f_i'(x, y)$ denotes the reconstructed shapes by Zernike moments from order 0 to i, then

$$f_i'(x, y) = F(|\sum_{n=0}^{i}\sum_{m} Z_{nm} V_{nm}(r, \theta)|) \qquad (13)$$

where F represents a mapping to gray level range 0-255, histogram equalization and threshold at 128.

Here we use Hamming distance $H(f_i', f)$ to measure the difference between the reconstructed shape f_i' and the original shape f_i. Pre-set a threshold ε, when $H(f_i', f) < \varepsilon$, we consider that ith order moments are enough to well represent the shape, and the higher orders are no longer calculated.

A feature vector is formed by the up to ith order Zernike moments, and then the weighted Euclidean distance is employed to match the features. The weight is determined by the following formula:

$$C(i) = H(f_{i-1}', f) - H(f_i', f) \qquad (14)$$

4   Experimental Results

4.1   Experiments on PSD

More than 200 shoeprint images are included in the experiment to evaluate the proposed algorithm. For each shoeprint, 5 partial images are generated to form the testing database, shown in Fig. 3. Examples of the testing database consist of 7 groups of shoeprint images are shown in Fig. 4 and the best matches can be found in table 1. Among the test data set of over 200 images, the recognition rate is around 86%.

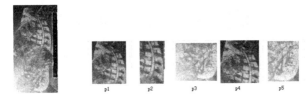

p1   p2   p3   p4   p5

Fig. 3. Partial shoeprints

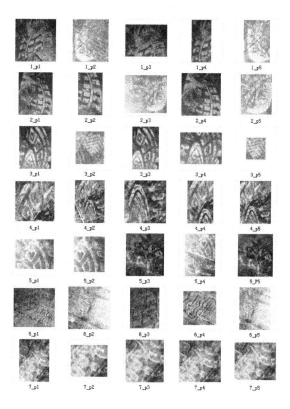

Fig. 4. Testing database

Table 1. Matching result of partial images in Fig.4

Group \ No.	p1	p2	p3	p4	p5
1	1_p3	1_p5	3_p1	3_p1	1_p2
2	2_p2	3_p3	2_p5	2_p1	2_p2
3	3_p3	3_p5	3_p1	3_p3	3_p2
4	4_p4	4_p5	4_p2	4_p1	4_p2
5	5_p2	5_p1	5_p5	5_p1	5_p3
6	6_p4	6_p5	3_p1	6_p1	6_p2
7	7_p5	7_p1	7_p1	7_p1	7_p1

4.2 Experiments on Zernike Moments

Since the shoeprint images taken from the crime scenes are always subjected to many distortions, a good shoeprint recognition method should have translation, scale and rotation invariance. For this purpose, an experiment for testing the invariance of

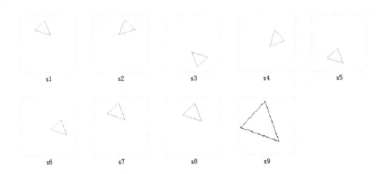

Fig. 5. Shapes for testing of invariance

Table 2. Testing results of invariance

Dis-tance	S1	S2	S3	S4	S5	S6	S7	S8	S9
S1	0	49.4	103.3	83.9	0	0	202.2	556.1	699.6
S2	49.4	0	83.0	75.2	49.4	49.4	211.1	574.3	691.3
S3	103.3	83.0	0	48.2	103.3	103.3	232.7	605.6	745.8
S4	83.9	75.2	48.2	0	83.9	83.9	223.0	585.7	746.9
S5	0	49.4	103.3	83.9	0	0	202.2	556.1	699.6
S6	0	49.4	103.3	83.9	0	0	202.2	556.1	699.6
S7	202.2	211.1	232.7	223.0	202.2	202.2	0	630.1	767.4
S8	556.1	574.3	605.6	585.7	556.1	556.1	630.1	0	432.7
S9	699.6	691.3	745.8	746.9	699.6	699.6	767.4	432.7	0

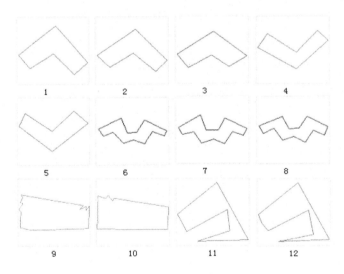

Fig. 6. A group of irregular shapes

Zernike moments has been firstly carried out. A set of testing shapes are generated from translation, rotation or scale transformation of an original shape, shown in Fig. 5. In table 2, s1 is the original shape, s2- s4 are the rotated versions of s1 while

Table 3. Distances between shapes 1-12 in Fig. 6

Distance (10^4)	1	2	3	4	5	6	7	8	9	10	11	12
1	0	0.23	0.41	0.45	0.32	0.76	0.68	0.76	0.69	0.70	0.61	0.59
2	0.23	0	0.22	0.32	0.18	0.61	0.53	0.60	0.64	0.66	0.67	0.63
3	0.41	0.22	0	0.38	0.30	0.50	0.43	0.48	0.52	0.58	0.80	0.76
4	0.45	0.32	0.38	0	0.14	0.55	0.49	0.55	0.70	0.71	0.69	0.61
5	0.32	0.18	0.30	0.14	0	0.61	0.53	0.60	0.67	0.68	0.67	0.60
6	0.76	0.61	0.50	0.55	0.61	0	0.11	0.06	0.65	0.63	0.91	0.84
7	0.68	0.53	0.43	0.49	0.53	0.11	0	0.11	0.61	0.57	0.89	0.81
8	0.76	0.60	0.48	0.55	0.60	0.06	0.11	0	0.63	0.61	0.89	0.82
9	0.69	0.64	0.52	0.70	0.67	0.65	0.61	0.63	0	0.13	1.03	0.99
10	0.70	0.66	0.58	0.71	0.68	0.63	0.57	0.61	0.13	0	1.06	1.01
11	0.61	0.67	0.80	0.69	0.67	0.91	0.89	0.89	1.03	1.06	0	0.24
12	0.59	0.63	0.76	0.61	0.60	0.84	0.81	0.82	0.99	1.01	0.24	0

s5-s6 and s7-s9 are translated and scaled shapes of s1 respectively. The data in table 2 shows that the distances between s1-s9 are relatively small which verifies the invariance of Zernike moments.

An experiment has been conducted to evaluate the discriminating ability of Zernike moments. More than 400 irregular shapes are randomly chosen from real shoeprint images to form the database. Fig. 6 shows a small subset of the extracted testing shapes. The difference between each of the two shapes in Fig. 6 is presented in table 3. Experimental data show that the distances between similar shapes are apparently smaller than the distances between unlike patterns, which proved Zernike moment to be powerful in shape discrimination. The recognition rate is around 98%.

5   Conclusion

In this paper, research has been done on shoeprint matching algorithms. The PSD value is calculated to match press down shoeprints and high quality spot shoeprints. In addition, considering the limitation of PSD method, we also provide a Zernike moments-based algorithm for shoeprint images with disordered background and blurred shoeprints. Compared with traditional techniques, our method is more applicable since it can match shoeprints with any complex sole patterns and has no requirements for the pre-processing of the image. Further more, it has a relatively low dependence on the operator.

References

1. Alexander, G.: Computer Classification of the Shoeprint of Burglar Soles. Forensic Science International 82, 59–65 (1996)
2. Sawyer, N.: 'SHOE FIT' A computerized shoeprint database. Institution of Electrical Engineers, UK (May 1995)
3. Geradts, Z., Keijzer, J.: The image data REBEZO for shoeprint with developments for automatic classification of shoe outsole designs. Forensic Science International 82, 21–31 (1996)

4. de Chazal, P., Reilly, R.B.: Automated Processing of Shoeprint Images Based on the Fourier Transform for Use in Forensic Science. IEEE Trans. on Pattern Analysis and Machine Intelligence 27, 341–350 (2005)

5. Huynh, C., de Chazal, P., McErlean, D., Reilly, R.B., Hannigan, T.J., Fleury, L.M.: Automatic classification of shoeprints for use in forensic science based on the Fourier transform, pp. 569–572. IEEE, Los Alamitos (2003)

6. Hu, M.K.: Visual pattern recognition by moment invariants. IRE Trans. on Info. Theory IT-8, 179–187 (1962)

7. Teague, M.R.: Image analysis via the general theory of moments. Journal of Optical Society of America 70, 920–930 (1980)

8. Teh, C.H., Chin, R.T.: On image analysis by the methods of moments. IEEE Trans. Pattern Analysis and Machine Intelligence 10, 496–513 (1988)

9. Khotanzad, A., Hong, Y.h.: Invariant Image Recognition by Zernike Moments. IEEE Trans. on Pattern Analysis and Machine Intelligence 12, 489–498 (1990)

10. Kim, W.S., Kim, Y.S.: A region-based shape descriptor using Zernike moments. Signal Process, Image Communication 16, 95–102 (2000)

11. Alexander, A., Bouridane, A., Crookes, D.: Automatic Classification and Recognition of Shoeprints. In: Proc. Seventh Int'l. Conf. Image Processing and Its Applications, vol. 2, pp. 638–641 (1999)

Shoe-Print Extraction from Latent Images Using CRF*

Veshnu Ramakrishnan, Manavender Malgireddy, and Sargur N. Srihari

University at Buffalo, State University of New York,
Amherst,
New York 14228
{vr42,mrm42,srihari}@cedar.buffalo.edu

Abstract. Shoeprints are one of the most commonly found evidences at crime scenes. A latent shoeprint is a photograph of the impressions made by a shoe on the surface of its contact. Latent shoeprints can be used for identification of suspects in a forensic case by narrowing down the search space. This is done by elimination of the type of shoe, by matching it against a set of known shoeprints (captured impressions of many different types of shoes on a chemical surface). Manual identification is laborious and hence the domain seeks automated methods. The critical step in automatic shoeprint identification is Shoeprint Extraction - defined as the problem of isolating the shoeprint foreground (impressions of the shoe) from the remaining elements (background and noise). We formulate this problem as a labeling problem as that of labeling different regions of a latent image as foreground (shoeprint) and background. The matching of these extracted shoeprints to the known prints largely depends on the quality of the extracted shoeprint from latent print. The labeling problem is naturally formulated as a machine learning task and in this paper we present an approach using Conditional Random Fields(CRFs) to solve this problem. The model exploits the inherent long range dependencies that exist in the latentprint and hence is more robust than approaches using neural networks and other binarization algorithms. A dataset comprising of 45 shoeprint images was carefully prepared to represent typical latent shoeprint images. Experimental results on this data set are promising and support our claims above.

1   Introduction

Approximately 30% of crime scene has usable shoeprints[2] left out by criminals. It is known that the majority of crime is committed by repeat offenders. It is common for burglars to commit a number of offences in the same day. The latent prints (figure 1(b)) which we get from the crime scene can be used for identification of suspects in a forensic case by narrowing down the search space.

As it would be unusual for an offender to discard their footwear between committing different crimes[1] timely identification and matching of shoe imprints

---

* Center of Excellence for Document Analysis and Recognition (CEDAR).

S.N. Srihari and K. Franke (Eds.): IWCF 2008, LNCS 5158, pp. 105–112, 2008.
© Springer-Verlag Berlin Heidelberg 2008

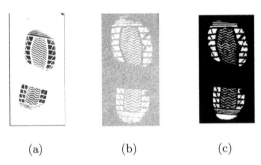

(a)   (b)   (c)

Fig. 1. Examples of different types of prints (a) chemical print, (b) latent print and (c) ground-truthed print. The image (c) is obtained by manually truthing the image (b) and is used for training and testing.

allows various crime scenes to be linked. So we resort help from automated methods. G. Alexandre [8] proposed a semi-automatic scheme for classifying shoes for burglars. Each sole is indexed with the type of patterns appearing in it e.g. circles, squares, zig zags, etc. A problem with this method is that the spacial location of the patterns are not coded and the shoe designs are becoming more complex making descriptions tedious. C. Huynh [7] proposed a solution for classifying shoes using fourier transform. Their database had known prints in which the extraction of the shoeprint is easier than from latent images. For extraction of shoe print from a latent print more complex work is required than just thresholding. A Bouridane [1] and Maria Pavlou [9] too came up with classification methods, but they too didn't handle the problem of extraction of shoeprint from latent images. The critical step in automatic shoeprint identification is *Shoeprint Extraction* because the print region has to be identified in order to compare it with the known print. An example of a known print is shown in figure 2. A known print is obtained in a controlled environment by using a chemical foot stamp pad. The impression thus formed are scanned. The shoeprint extraction problem can be formulated as a image labeling problem. Different regions(defined later in section 5) of a latent image are labeled as foreground (shoeprint) or background. The matching of these extracted shoeprints to the known prints largely depends on the quality of the extracted shoeprint from latent print. The labeling problem is naturally formulated as a machine learning task and in this paper we present an approach using Conditional Random Fields(CRFs). Similar approach was used for an analogous problem in handwriting labeling [3].The model exploits the inherent long range dependencies that exist in the latentprint and hence is more robust than approaches using neural networks and other binarization algorithms. Once the shoeprint has been extracted, the matching process is a task that is common to other forensic fields such as fingerprints where a set of features are extracted and compared. Our focus on this paper is only the shoeprint extraction phase and it suffices to say that the matching problem can be accomplished by using ideas from other forensic domains.

The rest of the paper is organized as follows. Section 2 describes our dataset and its acquisition. Section 3 described the CRF model followed by parameter estimation in section 4. Features for the CRF model are described in section 5 followed by experimental results and conclusion in section 7 and section 8 respectively.

2 Dataset

The two types of images that we require are the latent prints and the known(chemical) prints. The respective tasks of acquiring these are explained further below. A total of 45 latent and known print pairs are acquired.

2.1 Latent Prints

Bodziac [6] describes the process of recovery of latent prints. People are made to step on a powder and then on a carpet so that they leave their shoeprint on the carpet. Then the picture of the print is taken with a forensic scale near to the print. The current resolution of the image is calculated using the scale in the image and then it is scaled to 100dpi.

2.2 Known Prints

Chemical print is the known print which is obtained by a person stamping on a chemical pad and then on a chemical paper, which would leave clear print on a paper. All Chemical prints are scanned into images of resolution 100dpi.

2.3 Ground Truthed

The latent images are manually segmented by labeling the shoeprint as white and the background as black. These images (figure: 1(c)) are used to train and test the models.

3 Conditional Random Field Model Description

The probabilistic model of the Conditional Random Field used is given below.

$$P(\mathbf{y}|\mathbf{x},\theta) = \frac{e^{\psi(\mathbf{y},\mathbf{x};\theta)}}{\sum_{\mathbf{y}'} e^{\psi(\mathbf{y}',\mathbf{x};\theta)}} \tag{1}$$

where $y_i \in \{$Shoeprint, Background$\}$ and $\mathbf{x}$: Observed image and θ : CRF model parameters. It is assumed that a Image is segmented into 3X3 non-overlapping patches. The patch size is chosen to be small enough for high resolution and big enough to extract enough features. Then

$$\psi(\mathbf{y},x;\theta) = \sum_{j=1}^{m} \left(A(j,y_j,\mathbf{x};\theta^s) + \sum_{(j,k)\in E} I(j,k,y_j,y_k,\mathbf{x};\theta^t) \right) \tag{2}$$

The first term in equation 2 is called the state term(sometimes called Association potential as mentioned in [5]) and it associates the characteristics of that patch with its corresponding label. θ^s are called the state parameters for the CRF model. Analogous to it, the second term, captures the neighbor/contextual dependencies by associating pair wise interaction of the neighboring labels and the observed data(sometimes referred to as the interaction potential). θ^t are called the transition parameters of the CRF model. E is a set of edges that identify the neighbors of a patch. We use 24 neighborhood model. θ comprises of the state parameters,θ^s and the transition parameters,θ^t.
The association potential can be modeled as

$$A(j, y_j, \mathbf{x}; \theta^s) = \sum_i (f_i \cdot \theta_{ij})$$

where f_i is the ith state feature extracted for that patch and θ_{li} is the state parameter. The state features that are used for this problem are defined later in section 5 in table 1. The state features, f_l are transformed by the tanh function to give the feature vector $\mathbf{h}$. The transformed state feature vector can be thought analogous to the output at the hidden layer of a neural network. The state parameters θ^s are a union of the two sets of parameters θ^{s_1} and θ^{s_2}.

The interaction potential $I(\cdot)$ is generally an inner product between the transition parameters θ^t and the transition features f_t. The interaction potential is defined as follows:

$$I(j, k, y_j, y_k, \mathbf{x}; \theta^t) = \sum_l (f^l(j, k, y_j, y_k, \mathbf{x}) \cdot \theta_l^t)$$

4   Parameter Estimation

There are numerous ways to estimate the parameters of this CRF model [4]. In order to avoid the computation of the partition function we learn the parameters by maximizing the pseudo-likelihood of the documents, which is an approximation of the maximum likelihood value. For this paper, we estimate the Maximum pseudo-likelihood parameters using conjugate gradient descent with line search. The pseudo-likelihood estimate of the parameters, θ are given by equation 3

$$\hat{\theta_{ML}} \approx \arg\max_\theta \prod_{i=1}^M P(y_i|y_{\mathcal{N}_i}, \mathbf{x}, \theta) \tag{3}$$

where $P(y_i|y_{\mathcal{N}_i}, \mathbf{x}, \theta)$ (Probability of the label y_i for a particular patch i given the labels of its neighbors, $y_{\mathcal{N}_i}$), is given below.

$$P(y_i|y_{\mathcal{N}_i}, \mathbf{x}, \theta) = \frac{e^{\psi(y_i, \mathbf{x}; \theta)}}{\sum_a e^{\psi(y_i=a, \mathbf{x}; \theta)}} \tag{4}$$

where $\psi(y_i, x; \theta)$ is defined in equation 2.

Note that the equation 3 has an additional $y_{\mathcal{N}_i}$ in the conditioning set. This makes the factorization into products feasible as the set of neighbors for the patch form the minimal Markov blanket. It is also important to note that the resulting product only gives a pseudo-likelihood and not the true likelihood. The estimation of parameters which maximize the true likelihood may be very expensive and intractable for the problem at hand.

From equation 3 and 4, the log pseudo-likelihood of the data is

$$\mathcal{L}(\theta) = \sum_{i=1}^{M} \left(\psi(y_i = a, x; \theta) - log \sum_a e^{\psi(y_i = a, x; \theta)} \right)$$

Taking derivatives with respect to θ we get

$$\frac{\partial \mathcal{L}(\theta)}{\partial \theta} = \sum_{i=1}^{M} \frac{\partial \psi(y_i, x; \theta)}{\partial \theta} - \sum_a P(y_i = a | y_{\mathcal{N}_i}, x, \theta) \cdot \frac{\partial \psi(y_i = a, x; \theta)}{\partial \theta} \quad (5)$$

The derivatives with respect to the state and transition parameters are described below. The derivative with respect to parameters θ^{s2} corresponding to the transformed features $h_i(j, y_j, \mathbf{x})$ is given by

$$\frac{\partial \mathcal{L}(\theta)}{\partial \theta_{iu}} = \sum_{j=1}^{M} f_i(y_j = u) - \sum_a P(y_j = a | y_{\mathcal{N}_i}, x, \theta) f_i(a = u) \quad (6)$$

Here, $f_i(y_j = u) = f_i$ if the label of patch j is u otherwise $f_i(f_j = u) = 0$

Similarly, the derivative with respect to the transition parameters, θ_t is given by

$$\frac{\partial \mathcal{L}(\theta)}{\partial \theta_{l_{cd}}^{t}} = \sum_{j=1}^{M} \sum_{j,k \in E} f_l(y_j = c, y_k = d) - \sum_a \sum_{j,k \in E_{cd}} P(y_j = a | y_{\mathcal{N}_i}, x, \theta) f_l(a = c, y_k = d)$$

$$(7)$$

5   Features

Features of a shoeprint might vary according to the crime scene. It could be a powder on a carpet, mud on a table etc. So generalization of the texture of shoeprint is difficult. So we resort to the user to provide the texture samples of the foreground and background from the image. The sample size is fixed to be $15X15$

Table 1. Description of the 4 state features used

State Feature	Description
Entropy	Entropy of the patch
Standard Deviation	Standard deviation of the patch
Foreground Cosine Similarity	Cosine Similarity between the patch and the foreground sample
Background Cosine Similarity	Cosine Similarity between the patch and the background sample

Table 2. Description of the transition feature used - The transition feature is computed for a patch and its neighbor

Transition Feature	Description
Cosine Similarity	Cosine Similarity between the current patch and the surrounding 24 patches

which is big enough to extract information and small enough to cover the print region. There could be one or more samples of foreground and background. The feature vector of these samples are normalized image histograms. The two state features are the cosine similarity between the patch and the foreground sample feature vectors and the cosine similarity between the patch and the background sample feature vectors. The other two state features are entropy and standard deviation (table 1). The transition feature is the cosine similarity between the current patch and the surrounding 24 patches (table 2).

6   Inference

The goal of inference is to assign a label to each of the patches being considered. The algorithm for inference uses the idea of Gibbs sampling[10].

1. Randomly assign labels to each of the patches in a document based on an intuitive prior distribution of the labels.
2. Choose a patch at random and compute the probability of assigning each of the labels using the model from the equation given below to obtain a probability distribution p for the labels.

$$P(y_i|y_{N_i}, \mathbf{x}, \theta) = \frac{e^{\psi(y_i, \mathbf{x}; \theta)}}{\sum_a e^{\psi(y_i=a, \mathbf{x}\theta)}} \tag{8}$$

3. Use Gibbs sampling to sample from this distribution p to assign a probable label to the patch.
4. Repeat steps 2 and 3 until the assignments do not change. Store the set of label assignments along with the probability distribution p.
5. Repeat steps 1-4, for a sufficient number of iterations in order to eliminate the dependency on the initial random label assignments.
6. Consider the set of arrived assignments at step 4 in each of the iterations, and for all the patches pick the labels with the maximum probability as the final set of labels.

7   Experiments and Results

The shoeprint is converted into grayscale before processing. Segmentation is done using Otsu, Neural Network and Conditional Random Fields. Both Neural Network and Conditional Random Fields use the same feature set other than the

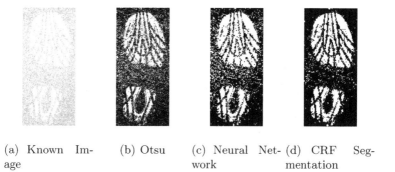

(a) Known Image   (b) Otsu   (c) Neural Network   (d) CRF Segmentation

Fig. 2. Segmentation results from different methods. The left most image is the input image and rest are the outputs from respective models. CRF seems to perform better the the previous two.

Table 3. Segmentation Results

Segmentation Method	Precision	Recall	F-Measure
Otsu	40.97	89.64	56.24
Neural Network	58.01	80.97	59.53
CRF	52.12	90.43	66.12

transition feature. The precision, Recall and F-measure is given in the table 3. These measures are calculated by comparing the result image with its respective ground truthed image (figure: 1(c)).

8 Conclusion

Performance of Otsu thresholding is not good if either the contrast between the foreground and the background is less or the background is non homogeneous. Neural Network seems to perform little better than the former by exploiting the texture samples that the user provided. CRF tend to outperform both by exploiting the neighborhood dependencies of the patches.

References

1. Bouridane, A., Alexander, A., Nibouche, M., Crookes, D.: Application Of Fractals To The Detection and Classification Of Shoeprints. IEEE, Los Alamitos (2000)
2. Alexandre, G.: Computerised classification of the shoeprints of burglars soles. Forensic Science International 82, 59–65 (1996)
3. Shetty, S., Srinivasan, H., Beal, M., Srihari, S.: Segmentation and labeling of documents using conditional random fields. In: Proceedings of SPIE, vol. 6500 (2007)
4. Wallach, H.: Efficient Training of Conditional Random Fields. In: Proc. 6th Annual CLUK Research Colloquium (2002)

5. Kumar, S., Hebert, M.: Discriminative Fields for Modeling Spatial Dependencies in Natural Images. Advances in Neural information processing systems (2003)
6. Bodziac, W.J.: Footwear Impression Evidence Detection, Recovery ond Examination. CRC Press, Boca Raton (2000)
7. Huynh, C., de Chazal, P., McErlean, D., Rei, R.B., Hannigan, T.J., Fleud, L.M.: Automatic Classification of Shoeprints for use in forensic science based on the fourier transform. In: ICIP (2003)
8. Alexandre, G.: Computer classification of the shoeprint of burglar soles. Forensic Science International 82, 59–65 (1996)
9. Pavlou, M., Allinson, N.M.: Automatic Extraction and Classification of Footwear Patterns. In: Corchado, E., Yin, H., Botti, V., Fyfe, C. (eds.) IDEAL 2006. LNCS, vol. 4224, pp. 721–728. Springer, Heidelberg (2006)
10. Casella, G., George, E.: Explaining the gibbs sampler. Amer. Statistician 46, 167–174 (1992)

Finding Identity Group "Fingerprints" in Documents

Lashon B. Booker

The MITRE Corporation, 7515 Colshire Drive,
McLean, Virginia 22102 USA
booker@mitre.org

Abstract. This paper describes how social identity group "fingerprints" can be extracted from a document collection by applying topic analysis methods in a novel way. The results of document classification experiments suggest that these group-level attributes provide better predictions of group affiliation than document-level attributes. Applications of this method for forensic authorship analysis are also discussed.

Keywords: topic analysis, latent dirichlet allocation, document classification, authorship analysis.

1 Introduction

The identity of an individual or target group responsible for authoring a text document or message can be a critical piece of evidence in many criminal investigations [10]. Computational approaches to authorship analysis usually focus on structural characteristics and linguistics patterns in a body of text [3]. While these approaches provide some important forensic capabilities, there remains a need for some way to discern the ideas and intentions conveyed in the text and use those qualities to help determine authorship [10].

Recently developed text analysis techniques may offer a feasible way to automatically compute such a semantic representation of text. Generative probabilistic models of text corpora, such as Latent Dirichlet Allocation, use mixtures of probabilistic "topics" to represent the semantic structure underlying a document [1]. Each topic is a probability distribution over words and the gist or theme of a document is represented as a probability distribution over those topics. Studies suggest that topic models give a better account of the properties of human semantic memory than latent semantic analysis models which represent each word as a single point in a semantic space [5].

When considering how to use this capability for authorship analysis, it is important to recognize that many factors influence the ideas present in a document, even when that document has just a single author. In particular, factors related to social identity - such as age, gender, ideology, beliefs, etc. – play an important role in communication behaviors. If the attributes of these factors could be teased out of a document, they might provide a valuable "fingerprint" facilitating author analysis. This paper describes preliminary experiments on a computational approach to extracting such identity group fingerprints.

S.N. Srihari and K. Franke (Eds.): IWCF 2008, LNCS 5158, pp. 113–121, 2008.
© Springer-Verlag Berlin Heidelberg 2008

Our investigation of these ideas begins by focusing on the role of identity groups in the formation of collaborative networks associated with scientific publications. This is a good starting point because many of the social factors influencing the content of a scientific publication can be readily identified and there are large amounts of publically available data to work with. We describe how to characterize identity groups and compute identity group attributes in this domain. We also present experiments showing that the group attributes provide better predictions of the contents of documents published by group members than attributes derived from the individual documents. Finally, we demonstrate how the techniques developed to extract identity group attributes can be used more broadly for document classification in general.

2 Identity Groups and Scientific Collaboration

An important goal of studies examining the collaborative networks associated with scientific publications is to understand how collaborative teams of co-authors are assembled. What are the important social identity groups that influence the way teams are assembled to write a paper in this setting? Publication venues are visible manifestations of some of those key social identity groups. People tend to publish papers in conferences where there is some strong relationship between their interests and the topics of the conference. Consequently, conference participants tend to have overlapping interests that give them a meaningful sense of social identity.

There are many attributes that might be useful for characterizing these social identity groups, ranging from the various social and academic relationships between individuals to the organizational attributes of professional societies and funding agencies. One readily available source of information about a group is the corpus of documents that have been collectively published by group members. Given such a document collection as a starting point, topics and frequently used keywords are an obvious choice as identity group attributes in this domain. The peer review system is a mechanism that ensures published papers include enough of the topics and keywords considered acceptable to the group as a whole. Authors that do not conform to these norms and standards have difficulty getting their papers published, and have difficulty finding funding for their work.

2.1 Topic Analysis

If topics are considered to be the identity group attributes of interest, it is natural to turn to topic analysis as a way of identifying the attributes characterizing each group. Topic analysis techniques have proven to be an effective way to extract semantic content from a corpus of text. Generative probabilistic models of text corpora, such as Latent Dirichlet Allocation (LDA), use mixtures of probabilistic "topics" to represent the semantic structure underlying a document [1]. Each topic is a probability distribution over the words in the corpus. The gist or theme of a topic is reflected in selected words having a relatively high probability of occurrence when that topic is prevalent in a document. Each document is represented as a probability distribution over the topics associated with the corpus. The more prevalent a topic is in a document, the higher its relative probability in that document's representation. We use an LDA model of the scientific document collections in our research.

An unsupervised Bayesian inference algorithm can be used to estimate the parameters of an LDA model; that is, to extract a set of topics from a document collection and estimate the topic distributions for the documents and the topic-word distributions defining each topic. As illustrated in Figure 1, the algorithm infers a set of latent variables (the topics) that factor the word document co-occurrence matrix. Probabilistic assignments of topics to word tokens are estimated by iterative sampling. See [4] for more details. Once the model parameters have been estimated, the topic distribution for a new document can be computed by running the inference algorithm with the topic-word distribution fixed (i.e., use the topic definitions that have already been learned).

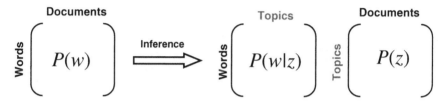

Fig. 1. Inferring the parameters of an LDA topic model

2.2 Document Classification

If topics are indeed identity group attributes for publication venues viewed as social identity groups, then these attributes ought to distinguish one group from another somehow. Different groups should have distinguishable topic profiles and the topic profile for a document should predict its group. Document classification experiments can be used to verify that identity group influence on published papers is reflected in document topic profiles.

The document descriptions derived from an LDA model are ideally suited to serve as example instances in a document classification problem. One outcome of estimating the parameters of an LDA model is that documents are represented as a probability distribution over topics. Each topic distribution can be interpreted as a set of normalized real-valued feature weights with the topics as the features. Results in the literature suggest that these features induced by an LDA model are as effective for document classification as using individual words as features, but with a big advantage in dimensionality reduction [1].

Regardless of what features we use, we would expect that documents from different topic areas would be distinguishable in a document classification experiment. A more interesting question is whether or not one set of features provides more discriminatory information than another. This is where our research hypothesis about group-level attributes becomes relevant. We hypothesize that in many cases the group identity is most effectively represented by group-level attributes. What are group-level attributes associated with the document collections being considered here?

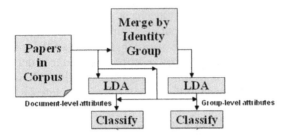

Fig. 2. Two ways to generate topic-based attributes for classifying documents

The standard approach to using LDA is to compute topics for the entire corpus that account for the word co-occurrences found in each individual document. We view these topics as document-level attributes. Our modeling emphasis on identity groups led us to consider an alternative approach that focuses instead on word co-occurrences at the group level. A very simple procedure appears to make this possible, as illustrated in Figure 2. First we aggregate the documents affiliated with an identity group into a single mega-document. Then, we use LDA to compute topics for the resulting collection of mega-documents. Since the topics computed in this way account for word co-occurrences in the mega-documents, they are attributes of the group and not attributes associated with the individual documents. However, these group-level topics can also be used as attributes to characterize each individual document, since all topics are represented as probability distributions over the same vocabulary of words. We hypothesize that the topics computed in this manner directly capture the attributes associated with each identity group as a whole and in some sense should be better than attributes derived from the word co-occurrences in individual documents.

2.3 Document Classification Experiments

In order to test this hypothesis, we conducted a document classification experiment in which topics extracted from an unlabeled body of text were used to predict the social identity group (that is, publication venue) of that document. The document collection for this experiment was selected from a dataset containing 614 papers published between 1974 and 2004 by 1036 unique authors in the field of Information Visualization [2]. The dataset did not include the full text for any of the documents, so we worked with the 429 documents in the dataset that have abstracts. The title, abstract and keywords of each entry constituted the "bag of words" to be analyzed for each individual document.

Several publication venues were represented here. We arbitrarily decided to consider each venue having 10 or more publications in the dataset as an identity group. This requirement provided some assurance that enough information was available to determine useful topic profiles for each group. Papers that did not belong to a venue meeting this minimum requirement were lumped together into a default group called "Other". Table 1 lists the identity groups and the number of documents associated with them. Given that the field of information visualization is itself a specialized identity group, it is not obvious that the smaller groups we specify here will have any

Table 1. The number of documents from each publication venue in the information visualization dataset

Source name	count
Proceedings of the IEEE Symposium on Information Visualization	152
IEEE Visualization	32
Lecture Notes in Computer Science (LNCS)	22
Conference on Human Factors in Computing Systems (SIGCHI)	21
ACM CSUR and Transactions (TOCS,TOID,TOG)	21
Symposium on User Interface Software and Technology (UIST)	17
IEEE Computer Graphics and Applications	13
IEEE Transactions	13
International Conference on Computer Graphics and Interactive Techniques (CGIT)	12
Communications of the ACM (CACM)	10
Advanced Visual Interfaces (AVI)	10
Other	106

distinguishable properties. There is a strong topic interdependency among these groups, which makes for a challenging classification task. It is also not obvious that broadly inclusive groups like the IEEE Symposium on Information Visualization or the "Other" category will have any distinguishable properties since they include papers from all the relevant topic areas in the field.

The selected documents were preprocessed to convert all words to lower case and remove all punctuation, single characters, and two-character words. We also excluded words on a standard "stop" list of words used in computational linguistics (e.g., numbers, function words like "the" and "of", etc.) and words that appeared fewer than five times in the corpus. Words were not stemmed, except to remove any trailing "s" after a consonant. This preprocessing resulted in a vocabulary of 1405 unique words and a total of 31,256 word tokens in the selected documents.

Two sets of topic features were computed from this collection: one from the set of individual documents and one from the set of aggregated mega-documents associated with each group. The optimal number of topics that fits the data well without overfitting was determined using a Bayesian method for solving model selection problems [4]. The model with 100 topics had the highest likelihood for the collection of mega-documents and the 200 topic model was best for the collection of individual documents. The resulting feature vector descriptions of each document were then used as examples for training classifiers that discriminate one identity group from another. Examples were generated by running the LDA parameter estimation algorithm over the words in a document for 500 iterations and drawing a sample of the topic distribution at the end of the run. We used a sparse representation for each example, only listing features which had a weight greater than the weight for a random choice. Ten examples were generated for each document[1], producing an overall total of 4290 examples for each feature set.

On each of ten runs of the experiment, a support vector machine classifier [6] was trained with a random subset of 75% of the examples, with the remaining 25% used for testing[2]. For each group, we trained a "one-versus-the-rest" binary classifier. This means that the examples from one class became positive examples while the examples from all other classes were treated as negative examples in a binary classification task. The overall solution to the multi-class problem is given by following the recommendation of the binary classifier with the highest classification score on a given test example. The support vector machine used a radial basis function kernel along with a cost factor to compensate for the unbalanced number of positive and negative examples. The cost factor weighs classification errors so that the total cost of false positives is roughly equal to the total cost of false negatives. For details about the cost factor, see [8].

Table 2. Results of the classification experiments on the information visualization dataset

Identity Group	# Examples	Document-level Accuracy	Group-level Accuracy
ACM CSUR	210	88.91%	96.91%
AVI	100	74.30%	98.06%
CACM	100	80.11%	91.77%
CGIT	120	82.01%	100.00%
IEEE Comp Graphics	130	80.87%	94.40%
IEEE Symp on InfoVis	1520	77.49%	87.34%
IEEE Transactions	130	71.81%	87.60%
IEEE Visualization	320	80.03%	90.94%
LNCS	220	95.40%	97.12%
SIGCHI	210	87.35%	90.06%
UIST	170	93.68%	90.73%
Other	1060	75.31%	83.89%
Overall	4290	79.74%	88.70%

The results of the document classification experiments are summarized in Table 2. Classification accuracy was computed by averaging over the ten independent runs. These results show that the group-level features produce a statistically significant improvement in overall classification accuracy over the document-level features on test data (88.7% accuracy versus 79.7% accuracy). Not surprisingly, the two most diverse classes ("IEEE Symposium on Information Visualization" and "Other") had

---

[1] Since the algorithm and representation are probabilistic, different runs will produce different feature vectors. The "true" feature vector can be thought of as a prototype that is most representative of all possible sample vectors.

[2] More specifically, we used 10-fold cross validation with each fold independently chosen, a method sometimes referred to as random subsampling.

the worst classification performance. The confusion matrix data shows that most classification errors were due to erroneous assignments to one of these classes.

These empirical results suggest that group-level attributes can provide better predictions of the contents of scientific documents published by group members than predictions based on attributes derived from the individual documents. This approach to document classification represents a modest step toward developing new approaches to modeling the effects of social identity groups on the behavior of individual members.

3 Authorship Analysis

Though the document classification approach based on topic analysis was designed to extract social identity group "fingerprints" from a document collection, it can be applied to a wide variety of document classification problems. The topic analysis does not depend on the existence of a social network of people who interact with each other[3]. The only requirement is that group or class labels are available for the document collection used for training.

In this section we show how our approach to document classification can be used in a setting that is relevant to forensic authorship analysis. We revisit a study [9] of how age and gender differences among bloggers are reflected in the writing style and content of blogs.

3.1 Age and Gender Effects on Blog Data

The Blog Authorship Corpus [9] was constructed using blogs collected from blogger.com in August 2004. Each blog selected for the corpus contained at least 500 words, including at least 200 occurrences of common English words, along with author-provided information about gender and age. From an initial collection of 46,947 blogs, a subset was extracted that included bloggers in three age categories: "10s" (ages 13-17), "20s" (ages 23-27), and "30+" (ages 33-47). Blogs in the "boundary" age groups 18-22 and 28-32 were removed in order to facilitate more reliable age categorization. Within each age category, the gender distribution was equalized by randomly discarding excess blogs from the larger gender group, leaving 8,240 "10s" blogs, 8,086 "20s" blogs and 2,994 "30+" blogs. The final corpus consists of 19,320 blogs containing 681,288 posts and over 140 million words (yielding approximately 35 posts and 7250 words per blog).

Schler *et al.* [9] used this corpus in a study that characterized differences in blogs based on style-related features and content-related features. Three types of style related features were considered: parts of speech (auxiliary verbs, pronouns, etc.); function words (frequently occurring English words such as "a", "it", "the", "very", etc.); and, "blogs words" – such as lol - and hyperlinks. The content-related features were simple content words that had the highest information gain among the frequently appearing words in a category. These content words were often closely associated with some distinct theme or topic. For example, the words "mother", "father", and "kids" are related to the theme "family".

---

[3] Moreover, the attributes shared by the group do not need to be linked to social identity.

In order to show that these vocabulary features could be used to predict the age and gender of a blog's author, Schler *et al.* constructed classification models for automated author profiling. Each blog was represented by a numeric vector whose entries were the frequencies, in that blog, of 502 style-related feature and 1000 content-related features. A linear-threshold machine learning algorithm [7] was applied to these vectors to generate classification models for author age and author gender. Empirical results show that these models can automatically classify unseen documents into the correct age category with an accuracy of 76.2% and identify the correct gender with an accuracy of 80.1%. This suggests that the vocabulary features apparently do capture important differences in writing style and content that distinguish bloggers with different genders and in different age categories.

3.2 Using Topic Analysis to Compute Age and Gender Attributes

Instead of trying to carefully select a subset of words as features that will discriminate between various age and gender categories, an intriguing alternative is to make automated feature discovery part of the authorship analysis problem. Features extracted using topic analysis techniques reflect a coupling between style and content as indicated by word co-occurrence patterns. This synergy may produce classification models with performance that is comparable to what can be obtained using hand-selected vocabulary features.

In order to test this hypothesis, we applied our topic analysis methodology to the Blog Authorship Corpus. We followed the same procedure used for the Information Visualization dataset, generating a model of 100 topics for the age group-level attributes and a separate model of 100 topics for the gender group-level attributes. These models were derived from a vocabulary of 148,201 unique words and a total of 26,048,869 word tokens in the corpus. Results show that these models can automatically classify unseen documents into the correct age category with an accuracy of 72.83% and identify the correct gender with an accuracy of 75.04%. This compares favorably with the results reported by Schler *et al.* that were based on a much larger number of hand-selected features.

4 Summary

This paper has shown how identity group "fingerprints" can be extracted from a document collection by applying topic analysis methods in a novel way. Empirical results on scientific publication data suggest that group-level attributes can provide better predictions of the contents of scientific documents published by group members than predictions based on attributes derived from the individual documents. This argues in favor of using group-level attributes for document classification.

Experiments with blog data show that this document classification method can also be effective for forensic authorship analysis tasks. Besides providing good classification accuracy, it has the added benefit of automatically inferring a good set of features that appear to account for both style-related and content-related differences in vocabulary usage. This capability could be a useful supplement to forensic methods that incorporate other techniques such as linguistics and behavioral profiling.

Acknowledgments. This research was funded by the MITRE Sponsored Research program. That support is gratefully acknowledged.

References

1. Blei, D., Ng, A., Jordan, M.: Latent Dirichlet Allocation. Journal of Machine Learning Research 3, 993–1022 (2003)
2. Börner, K., Dall'asta, L., Ke, W., Vespignani, A.: Studying the Emerging Global Brain: Analyzing and Visualizing the Impact of Co-Authorship Teams. Complexity 10(4), 57–67 (2005)
3. De Vel, O., Anderson, A., Corney, M., Mohay, G.: Mining E-mail Content for Author Identification Forensics. SIMOD Record 30(4), 55–64 (2001)
4. Griffiths, T., Steyvers, M.: Finding scientific topics. Proceedings of the National Academy of Science 101, 5228–5235 (2004)
5. Griffiths, T., Steyvers, M., Tenenbaum, J.: Topics in semantic representation. Psychological Review 114(2), 211–244 (2007)
6. Joachims, T.: Making large-scale SVM learning practical. In: Schölkopf, B., Burges, C., Smola, A. (eds.) Advances in Kernel Methods – Support Vector Learning. MIT Press, Cambridge (1999)
7. Littlestone, N.: Learning quickly when irrelevant attributes abound: A new linear–threshold algorithm. Machine Learning 2(4), 285–318 (1988)
8. Morik, K., Brockhausen, P., Joachims, T.: Combining statistical learning with a knowledge-based approach – A case study in intensive care monitoring. In: Proceedings of the 16th International Conference on Machine Learning, pp. 268–277. Morgan Kaufmann, San Francisco (1999)
9. Schler, J., Koppel, M., Argamon, S., Pennebaker, J.: Effects of Age and Gender on Blogging. In: Computational Approaches to Analyzing Weblogs: Papers from the 2006 AAAI Spring Symposium. AAAI Press, Menlo Park (2006)
10. Stokar von Neuforn, D., Franke, K.: Reading Between the Lines: Human-centered Classification of Communication Patterns and Intentions. In: Liu, H., Salerno, J., Young, M. (eds.) Social Computing, Behavioral Modeling, and Prediction. Springer, New York (2008)

Supporting Law Enforcement in Digital Communities through Natural Language Analysis

Danny Hughes[1], Paul Rayson[1], James Walkerdine[1], Kevin Lee[2], Phil Greenwood[1],
Awais Rashid[1], Corinne May-Chahal[3], and Margaret Brennan[4]

[1] Computing, InfoLab 21, South Drive, Lancaster University, Lancaster, UK, LA1 4WA
{danny, paul, walkerdi, greenwop, marash}@comp.lancs.ac.uk
[2] Isis Forensics, P.O. Box 793, Lancaster, LA1 9ED, UK
k.lee@isis-forensics.com
[3] Department of Applied Social Science, Lancaster University, Lancaster, UK, LA1 4YL
c.may-chahal@lancs.ac.uk
[4] Child Exploitation and Online Protection Centre, 33 Vauxhall Bridge Road,
London, UK, SW1V 2WG
maggie.brennan@ceop.gov.uk

Abstract. Recent years have seen an explosion in the number and scale of digital communities (e.g. peer-to-peer file sharing systems, chat applications and social networking sites). Unfortunately, digital communities are host to significant criminal activity including copyright infringement, identity theft and child sexual abuse. Combating this growing level of crime is problematic due to the ever increasing scale of today's digital communities. This paper presents an approach to provide automated support for the detection of child sexual abuse related activities in digital communities. Specifically, we analyze the characteristics of child sexual abuse media distribution in P2P file sharing networks and carry out an exploratory study to show that corpus-based natural language analysis may be used to automate the detection of this activity. We then give an overview of how this approach can be extended to police chat and social networking communities.

Keywords: Social Networks; P2P; Network Monitoring; Natural Language Analysis; Child Protection.

1 Introduction

Social networking sites, chat applications and peer-to-peer (P2P) file sharing systems support millions of users and have redefined how people interact on the Internet. Millions of people use social networking sites such as MySpace [1] and chat applications such as MSN messenger [2] to support their personal and professional communications creating so-called *digital communities*. Similar communities have been created in P2P file sharing systems, such as Gnutella [3], which enables the decentralised sharing of files free from control or censorship by third parties.

As the scale of digital communities and their importance to society grows, there is a strong tendency for them to reflect real communities with the infiltration of criminal

S.N. Srihari and K. Franke (Eds.): IWCF 2008, LNCS 5158, pp. 122–134, 2008.
© Springer-Verlag Berlin Heidelberg 2008

activities. For example, social networking sites have given rise to the phenomenon of 'cyber-stalking' [4] while chat applications have been used by paedophiles to support online victimisation, such as the 'grooming' of children [5]. Similarly, P2P file sharing systems have been implicated in both copyright infringement and the distribution of material related to child sexual abuse [6]. Digital communities, therefore, present two major and related challenges to law enforcement:

1. Digital communities provide new and easier ways for criminals to organise. An example of this is the ability of paedophiles to formulate ad-hoc networks to exchange child sexual abuse media and, even more seriously, to plan child sexual abuse activities.
2. The extremely large scale of digital communities coupled with rapidly evolving underlying protocols renders pro-active manual policing (analogous to police patrols in the physical world) infeasible. This can leave children and vulnerable adults exposed to significant risk.

This paper focuses on advancing methods for safeguarding children in digital communities and argues that effective pro-active policing of digital communities requires the use of automated language analysis techniques. These techniques, inspired by the fields of computational and corpus-based linguistics, can be used to help law enforcement agencies to identify:

i. Criminal activity.
ii. Evolving and emergent criminal terminology.
iii. Child predators masquerading as children.

To illustrate the scale of this problem, we perform an analysis of the scale and characteristics of the distribution of child sexual abuse media on P2P file sharing systems. Our study reuses tracing data collected and reported on in a previous study [6], however the analysis reported here is new and focuses specifically on child sexual abuse. We then describe an exploratory study involving the classification of emerging criminal terminology using natural language analysis. Finally, we discuss how this approach could be extended to support the policing of chat and social networking sites in order to prevent paedophiles using these systems to victimise children. While, in this paper, we focus on tackling paedophilia in digital communities, the techniques can be expanded to other criminal activities, for instance, to identify terrorists recruiting impressionable youth or organising attacks through digital communities.

The remainder of this paper is structured as follows: section 2 provides an overview of the prevalent digital communities of today. Section 3 introduces the background to our natural language analysis approach and its application to the policing of digital communities. Section 4 presents an analysis of the scale and characteristics of the distribution of child-abuse media on P2P file sharing systems. Section 5 evaluates the effectiveness of natural language analysis to automate the identification of child-abuse media. Section 6 provides a discussion on how natural language analysis can also be used to support the pro-active policing of chat and social networking sites for various criminal activities, and section 7 concludes and highlights areas of future work.

2 Digital Communities

Today's major digital communities include: P2P file sharing systems, chat applications and social networking sites. We first review the existing evidence of criminal activity in these communities and subsequently highlight the policing challenges.

2.1 P2P File Sharing

Peer-to-peer file sharing systems use the bandwidth and disk space of home or office computers to maintain a distributed online library of files. Any P2P user may directly share files with any other user, in a largely anonymous and censorship-resistant way. P2P has revolutionised how people use the Internet to communicate, empowering users to produce and distribute content free from any form of control or censorship by third parties. Today it is widely accepted that P2P applications (for example, Gnutella [3] and Bittorrent [7]) are responsible for the vast majority of internet traffic [8].

Unfortunately, this perceived anonymity has led to wide-spread illegal behaviour, most notably the use of P2P file sharing systems to distribute illegal content. Current research suggests that 90% of all material on P2P file sharing networks is copyrighted [9]. Likewise a recent study [6] showed that 1.6% of searches and 2.4% of search-responses on the Gnutella P2P network relate to illegal sexual content such as images of child sexual abuse. These levels equate to hundreds of searches for illegal images per second. While this work illustrates the significant scale of this problem, it also shows that the distributors of such material tend to form sub-communities, which do not interact with the broader P2P community.

2.2 Chat Applications

In a similar vein to P2P, chat applications have had a dramatic impact on how people communicate. Although chat software has existed for decades in guises such as IRC [10], it is only within the last ten years that chat applications have become adopted by mainstream Internet users. This is perhaps best illustrated by Instant Messenger applications such as MSN [2] and Skype [11] that are now common tools in both the home and workplace.

As chat applications have become popular, they have also become a means to support criminal activities. Their growing use by children and young people has given paedophiles a new means to target children (for example, [12]), or even to locate other paedophiles and plan paedophilia-related activities (for example, [13]).

2.3 Social Networking Sites

Social networking sites such as MySpace [1] and Facebook [17] are the newest form of digital community to be adopted by mainstream computer users, and like chat and P2P file sharing applications, participation is expanding quickly— in June 2007 "Social Media Today" reported that Facebook alone had 25 million users [18]. As of April 2008, Facebook claims to support 70 million users [34]. As with chat and P2P systems, the growth of this type of digital community has created opportunities for criminal behaviour. Users of social networking sites are particularly vulnerable as these sites allow individuals to rapidly expand their social network to include un-vetted strangers. Furthermore, the tendency of some users to post personal information such as their real-world address, school/job and telephone number allows cyber-criminals unprecedented access to potential victims. In particular, this has been exploited by cyber-stalkers [19], who use personal information along with the access that social networking sites provide to harass their victims (for example [20]). While cyber-stalking has been recognised as an offence that can be prosecuted under a range of existing legislation (such as the UK Protection from Harassment Act, 1997), difficulties arise from the variable range of legislation in countries across the world. Social networking sites are also used by paedophiles as another avenue through which to approach children. For example, paedophiles have been observed masquerading as children in order to initiate contact with their victims [21]. While there is a need to educate users of social networking sites about protecting their personal data, there is also an urgent need for effective policing of these communities to protect vulnerable users, for example, children.

2.4 Problems Inherent in Policing Digital Communities

Efforts to combat the concerns highlighted above have been reflected by the formation of the Virtual Global Taskforce (VGT) [14], the launch of organisations such as the Child Exploitation and Online Protection centre (CEOP) in the UK [15], and the introduction of legislation to criminalise the 'grooming' of children in chat rooms [16]. However, law enforcement agencies policing digital communities in order to detect, prevent and prosecute criminal activity on these systems face three major challenges:

- *Dealing with large volumes of data*: With millions of users, digital communities generate vast amounts of data, which makes manual analysis infeasible.
- *Identifying criminal activity*: Criminals such as paedophiles often develop their own vocabulary of terms to disguise their activities.
- *Identifying masquerading users*: In most digital communities, the creation of fake user identities is trivial, allowing criminals to evade detection (e.g. paedophiles posing as a children to contact their victims).

Critically, the very characteristic of these digital communities - that is, the lack of routine feedback such as body language, tone of voice or facial expressions [4], hampers policing and inevitably puts greater emphasis on language use, which can, in turn, be used to aid policing. Specifically, we propose that internet policing problems may be tackled more effectively through automated natural language analysis of traffic originating from digital communities.

Such an approach can identify likely criminal activity amongst the huge volumes of innocent interactions occurring in digital communities and furthermore, through the use of language profiling, make it possible to detect paedophiles pretending to be children and other masquerading users (such as cyber-stalkers using fake personas). A brief overview of our natural language analysis approach and its application to detecting and preventing crime in digital communities is provided next. For more details on the set of natural language tools and techniques we deploy for the purpose, interested readers are referred to [33].

3   Natural Language Analysis

Existing work on policing online social networks has focused primarily on the monitoring of chat and file sharing systems. Chat policing software for home use such as Spector Pro [29], Crisp [30] and SpyAgent [31] allow the logging of online conversations, but are restricted in that they need to be installed on the actual PC that is being policed. In terms of language monitoring capabilities, the existing chat policing software tools rely on human monitoring of logs or simple-minded word or phrase detection based on user-defined lists. Such techniques do not scale. Nor do they enable identification of adults masquerading as children or support pro-active policing.

Techniques do exist which make use of statistical methods from computational linguistics and corpus-based natural language processing to explore differences in language vocabulary and style related to the age of the speaker or writer. Keyword profiling [22], exploiting comparative word frequencies, has been used in the past to investigate the differences between spoken and written language [23], British and American English [24], and language change over time. Rayson [25] extended the keywords methodology to extract key grammatical categories and key domain concepts using tagged data in order to make it scalable. The existing methodologies draw on large bodies of naturally occurring language data known as corpora (sing. corpus). These techniques already have high accuracy and are robust across a number of domains (topics) and registers (spoken and written language) but have not been applied until now to uncover deliberate deception. The second relevant technique is that of authorship attribution [26]. The current methods [27] would allow a narrowing in focus from the text to the individual writer in order to generate a stylistic fingerprint for authors.

Our particular focus in this paper is file sharing systems where filenames and search terms reflect specialised vocabulary which changes over time. Using a frequency profiling technique, we can find popular terms within the search query corpus. Known terms such as high frequency words normally expected within a general language corpus (e.g. the, of, and, a, in) can be eliminated either manually or by using a list of 'stop words', a technique often used in information retrieval. From this list emergent unclassified terminology may be identified, and through association classified, allowing for the detection of emergent criminal terminology.

Section 4 first explores the scale and characteristics of the distribution of child sexual abuse related media on P2P file sharing networks. Section 5 then explores the use of natural language analysis to terminology emerging from this deviant sub-community.

4   Child Sexual Abuse Media Distribution on File Sharing Networks

Our previous research has shown that P2P file sharing networks, and specifically Gnutella, are a major vector for the distribution of illegal sexual material [6]. We now revisit our experiments on the Gnutella P2P file sharing network in order to specifically analyze the extent to which this system is used to distribute media related to child sexual abuse. Section 4.1 presents our experimental methodology and section 4.2 presents our results.

4.1 Experimental Methodology

Gnutella is an open protocol designed to support the discovery and transfer of files among its users. In technical terms, the Gnutella protocol builds an unstructured, decentralized network where peers are required to forward network maintenance and file discovery messages, and to share files on the network. File discovery on Gnutella uses two message types. 'QUERY' search messages that are broadcast on the network with an XML or plain text search term to discover files. Plain text/XML 'QUERY-HIT' search-response messages are used to inform a searching peer that a matching file is available. Thus by connecting to the Gnutella network and intercepting and analyzing QUERY and QUERYHIT messages we can explore what files users are searching for, and offering respectively.

In our previous experiment to quantify the level of resource discovery traffic that relates to illegal sexual material [6] we gathered a one month trace of Gnutella traffic (between February 27th and March 27th 2005). Two independent reviewers then analysed 10,000 search and search-response messages from three separate days within our trace (5th, 12th and 19th March), classifying them as relating to either illegal sexual material or other material. Our approach was to classify messages as relating to illegal pornographic material if they could only be interpreted as referring textually to such material. We found that an average of 1.6% of searches and 2.4% of search responses contained references to such material.

We have since revisited this data in order to analyze the level of resource discovery traffic that relates specifically to *child sexual abuse*. Once again, two independent reviewers analyzed three samples of 10,000 search and search-response text messages from the 5th, 12th and 19th March 2005. The results of our experiments are provided in section 4.2. As with our previous experiments, the understanding that can be provided through an analysis of meta-data is limited – while our results do show that the level of resource discovery traffic relating to illegal sexual material, there is no way to know whether these searches were generated deliberately or in error. Nor is it possible to know whether a user performing such a search found and/or downloaded the corresponding material. Thus the analysis of resource discovery traffic provides a useful, but imperfect measure of the extent to which P2P file sharing networks are used to distribute material relating to child sexual abuse. Section 4.2 presents the results of our experiments.

4.2 Level of Child Sexual Abuse Related Resource Discovery Traffic on Gnutella

The results of the two independent reviewers' traffic classifications are shown in tables 1a and 1b:

Table 1a. Reviewer 1, Child-Abuse Related Resource Discovery Traffic

	5th March	12th March	19th March
Search Messages	0.9% (90)	1.55% (155)	0.81% (81)
Search Responses	1.59% (159)	1.83% (183)	1.25% (125)

Table 1b. Reviewer 2, Child-Abuse Related Resource Discovery Traffic

	5th March	12th March	19th March
Search Messages	0.90% (90)	1.4% (140)	0.86% (86)
Search Responses	1.59% (159)	1.9% (190)	1.38% (138)

It can be seen that there is a high degree of correlation between the independent reviewers' classifications:

- An average of 1.07% of the search message sample was classified as relating to child sexual abuse. The minimum value we observed was 0.81% on March 19th, rising to 1.55% on March 12th. The standard deviation between samples was 0.32%.
- An average of 1.59% of the search responses relate to child sexual abuse. The minimum value observed was 1.25% on March 19th, rising to 1.9% on March 12th. The standard deviation was 0.25%. The higher level of search responses that contain references to child sexual abuse arises because search responses list multiple files.

Given the large scale of the Gnutella network, this equates to thousands of child-abuse related searches and search responses per minute, suggesting that P2P file sharing systems and Gnutella specifically are a major vector for this material. Unfortunately, a major problem in identifying the presence of child-abuse related media is the use of domain-specific terminology by paedophiles. Our trace data was further analysed by domain experts for the presence of non-standard child abuse terminology. We found that, on average 53% of searches and 88% of responses used such language. Section 5 discusses and evaluates the use of natural language analysis to identify such terminology emerging from criminal sub-communities.

5 Identifying Child Sexual Abuse Media Using Natural Language Analysis

This section introduces a simple scheme to detect and classify emerging terminology related to the distribution of child sexual abuse media. Section 5.1 describes our experimental methodology. Section 5.2 reports on our results. Finally section 5.3 discusses how this process may be fully automated.

5.1 Experimental Methodology

These experiments were performed using our Gnutella trace data [6] and assess the viability of using *association* to classify previously unknown terminology relating to child sexual abuse media. The classification process was performed in a semi-automated fashion by 10 volunteers with no domain-specific knowledge (i.e. links to child-protection services, law-enforcement agencies or related criminal convictions). Our hypothesis is that by identifying non-standard search words that have a high popularity and providing context by showing complete search phrases that contain these words, we can significantly improve the accuracy of manual catagorisation. A complete list of the search words being categorized can be obtained by emailing the authors (due to their sensitive nature they are not reproduced here).

Firstly, our natural language analysis tool (Wmatrix [28]) was used to create a frequency-ranked list of the 1000 most popular search words. From this list, the top 5 non-standard language search words were extracted for non-child sexual abuse related material and the top 5 non-standard language words were extracted for child sexual abuse related material. Each volunteer in our test-group was then asked to classify these words as either (i) related to child sexual abuse or (ii) unrelated to child sexual abuse.

Following this 'blind' classification of terminology, each volunteer took part in a semi-automated process wherein association with complete search phrases was used to suggest the meaning of the unknown search terms. Specifically, each volunteer was presented with 10 complete search phrases containing the unknown term that were selected at random from our trace data. Each volunteer then reclassified every term based on its appearance with known (English) language words contained in the search phrases presented. Section 5.2 describes the results of this experiment and section 5.3 describes how the process may be further automated in order to improve efficiency and scalability.

5.2 Experimental Results

Table 2 shows the results of the initial blind catagorisation and subsequent catagorisation via association with complete search terms. It can be seen that blind catagorisation of child-abuse related media had an average success rate of only 42%, while classification through association more than doubled the success rate to 94%. Firstly, this shows that natural language analysis can provide significant benefits for identifying emergent terms. Secondly, it illustrates that the association with previously classified language is a promising approach to supporting the classification of new terminology.

Table 2. Use of Association to Improve Classification of Child Sexual Abuse Related Terms

	Anonymous Reviewer									
	1	2	3	4	5	6	7	8	9	10
Success of blind catagorisation (%).	40	40	20	60	60	0	40	40	80	40
Success of catagorisation with association (%).	80	100	100	100	100	80	100	100	100	80

The results of this experiment are shown graphically in Figure 1 below. It can be seen that the ability of test subjects to blindly classify child sexual abuse related terminology varied considerably - from 0% for reviewer 6 to 80% for reviewer 9. In all cases the use of association improved the subjects' ability to correctly classify terms – by an average of 52%.

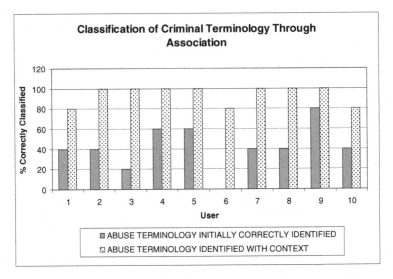

Fig. 1. Use of Association to Classify Child Sexual Abuse Related Terms

5.3 Automating the Detection and Classification of New Terminology

In order to automate the detection of new terminology, we can supplement the frequency profiling technique above with *comparative frequency analysis*. This technique compares the frequency profile generated from search terms to a frequency profile built from a reference corpus of general English. For each word in the profile, a log-likelihood (LL) statistical test is performed which estimates the significance of the difference in its frequency between the search term corpus and the reference corpus [32]. A larger LL value indicates that the difference in relative frequencies observed is larger and less likely to occur by chance. Therefore by sorting the resulting comparative profile on the LL value, we could find the most overused words in the search term corpus relative to an English standard reference. These key words appearing at the top of the list are most likely to be domain-specific words. This technique could be further extended by maintaining a *monitor corpus* of search terms consisting of trace data over the last 24 hours or 7 days, and using this as a reference corpus instead of the general English dataset. This would improve the extraction of novel terminology and enable the monitoring of short-term trends in search terms.

For classification of new terminology, we can exploit computer techniques that mirror the activities of the human volunteers during in the second stage of our xperiment. The extra information provided by viewing example terms in context provides evidence for the reader to make an educated guess at the meaning of an unknown term. A technique from corpus linguistics called *collocation* provides a way of calculating the strength of association between one word and others in the surrounding context. Pairs of words with high collocation scores occur significantly more often together than would be expected by chance based on their individual frequencies. Thus, if we find that an unknown word regularly collocates with other words that are known to be related to child-abuse media, we can make the assumption that the unknown word is also specific to that domain. This has the potential to more quickly identify emerging terminology and enhance the effectiveness of policing.

6 Discussion: Applying Natural Language Analysis to Chat and Social Networking Systems

The case for supporting the policing of chat and social networking systems is even more compelling than the case for supporting the policing of P2P file sharing systems. While P2P file sharing systems are a major vector for the distribution of existing media relating to child sexual abuse, chat and social networking systems are actively used to support the perpetrators of abuse. As such, supporting the pro-active policing of these systems is critical to preventing future crimes.

Fortunately, natural language analysis also holds significant promise for supporting the policing of chat and social networking systems. Domain-specific language relating to child abuse is largely unused except by those involved with the distribution of child abuse related materials. Thus, the detection of this language in chat rooms or on social-networking sites (e.g. its use by paedophiles planning a crime) may be used by law enforcement bodies to better target their limited manual policing resources.

In cases where paedophiles are masquerading as children on social networking or chat systems (e.g. to gain access to a child), the use of domain specific terminology is unlikely. In such cases, natural language analysis may still be employed to discover masquerading paedophiles. This requires the establishment and extension of corpora of child and adult language in chat rooms. Comparing samples of observed chat language with these corpora using the techniques described in section 3 may provide some clues to adults masquerading as children in terms of their vocabulary usage. However, we expect to have to extend these techniques to grammatical profiles in order to identify stylistic clues to deceptive language.

In either case the monitoring of digital communities also necessitates the development of new tracing functionality to support the establishment and extension of necessary corpora along with policing. For those digital communities which utilise peer-to-peer communication protocols, our existing monitoring approaches can be readily adapted [6]. However, those systems which use a more centralised communications infrastructure will necessitate the investigation of approaches similar to those used to implement web-based data mining [35].

7 Conclusions and Future Work

This paper has argued that natural language analysis is a promising approach to support the policing of digital communities. To illustrate the critical need for policing support, we first analysed the scale and characteristics of the distribution of child sexual abuse media on the Gnutella P2P file sharing network, finding that more than 1% of search messages and 1.5% of search-response messages relate to this material. We then showed that natural language analysis can be used in a semi-automated methodology to identify non-standard language used by those distributing child abuse media. Finally, we discussed how such an approach can be automated and applied across other digital communities such as chat and social networking systems.

The work presented here uses one of a number of techniques from corpus-based linguistics that we intend to trial in our approach to supporting the detection and classification of unknown vocabulary in chat and social networking systems. In future work we aim to: (a) to integrate the statistically sophisticated but knowledge-poor techniques from authorship attribution (that operate on one person's language) with linguistically-informed approaches in key domain analysis (that operate at the level of language of groups), and (b) to develop novel methods that are capable of detecting adults masquerading as children within the constraints of small amounts of language evidence, in the region of hundreds rather than thousands of words, that are observed in chat room data. Detection of unknown vocabulary is a relatively straightforward issue (comparable to a spell checker using a full form dictionary), but the task of classification of the unknown terms will require novel extension of techniques such as unsupervised word sense disambiguation where we will investigate exploitation of regular collocation with known domain vocabulary to seed the technique.

Any form of monitoring also raises ethical considerations. When extending and applying our monitoring approach to digital communities, such ethical issues will also be considered. This is important not only in respect of the abuse of children and the protection of the system developers, but also to the privacy of digital community members. The level of monitoring needs to be carefully and clearly explained to digital community members for them to accept the proposed approach. Part of the research performed will involve examining ethical issues and ensuring such issues are addressed throughout the development life-cycle of the monitoring approach.

Acknowledgements

This research is supported by a research grant from EPSRC/ESRC (reference EP/F035438/1) involving the Universities of Middlesex and Swansea as well as the Child Exploitation and Online Protection Centre (CEOP), project title "Isis: Protecting children in online social networks". We would also like to thank Isis Forensics (www.isis-forensics.com) for helping to conduct the experiments.

References

[1] MySpace (April 2008), http://www.myspace.com/
[2] MSN Messenger (April 2008), http://webmessenger.msn.com/
[3] The Gnutella Protocol Specification, version 0.4 (retrieved, April 2008),
    http://www9.limewire.com/developer/gnutella_protocol0.4.pdf
[4] Ellison, L.: Cyberstalking: Tackling Harassment on the Internet. In: 14th BILETA Conference: CYBERSPACE 1999: Crime, Criminal Justice and the Internet (1999)

[5] Pallister, D.: Internet paedophile gets nine years for sex with schoolgirls, Guardian Newspaper (June 23, 2006), http://www.guardian.co.uk/uk/2006/jun/23/ukcrime.davidpallister

[6] Hughes, D., Gibson, S., Walkerdine, J., Coulson, G.: Is Deviant Behaviour the Norm on P2P File Sharing Networks? IEEE Distributed Systems Online 7(2) (February 2006)

[7] Bittorrent Protocol Specification, version 1.0 (retrieved, April 2008), http://cs.ecs.baylor.edu/~donahoo/classes/5321/projects/bittorrent/BitTorrent%20Protocol%20Specification.doc

[8] Karagiannis, T., Broido, A., Brownlee, N., Faloutsos, M.: Is P2P Dying or Just Hiding? In: Proceedings of Globecom 2004, Dallas, Texas, USA (December 2004)

[9] Lee, K., Walkerdine, J., Hughes, D.: On the Penetration of Business Networks by P2P File Sharing. In: Proceedings of the 2nd International Conference on Internet Monitoring and Protection (ICIMP 2007), Santa Clara, California, USA (July 2007)

[10] RFC 1459: Internet Relay Chat (IRC) Protocol (retrieved, April 2008), http://www.irchelp.org/irchelp/rfc/

[11] Skype (April 2008), http://www.skype.com

[12] BBC News 24, Chat room Paedophile Jailed, http://news.bbc.co.uk/1/hi/england/2969020.stm

[13] BBC News 24, Men Jailed for Online Rape Plot (April 2008), http://news.bbc.co.uk/1/hi/england/6331517.stm

[14] The Virtual Global Task Force (April 2008), http://www.virtualglobaltaskforce.com/

[15] The UK Child Exploitation and Online Protection Centre (CEOP) (April 2008), http://www.ceop.gov.uk

[16] Scottish Parliament, The Protection of Children and Prevention of Sexual Offences (Scotland) Bill (April 2008), http://www.scottish.parliament.uk/business/bills/pdfs/b30s2-aspassed.pdf

[17] Facebook (April 2008), http://www.facebook.com

[18] Social Media Today, Facebook Explodes (June 2007), http://www.socialmediatoday.com/SMC/10670

[19] Office of Public Sector Information, Malicious Communications Act 1988 (April 2008), http://www.opsi.gov.uk/ACTS/acts1988/Ukpga_19880027_en_1.htm

[20] Crime Library (2007), Cyberstalking- A Case Study (April 2008), http://www.crimelibrary.com/criminal_mind/psychology/cyberstalking/5.html

[21] Panorama Transcript: One click from Danger (2008) (April 2008), http://news.bbc.co.uk/1/hi/programmes/panorama/7180769.stm

[22] Scott, M.: Focusing on the text and its key words. In: Burnard, L., McEnery, T. (eds.) Rethinking Language Pedagogy from a Corpus Perspective, Peter Lang, Frankfurt, pp. 104–121 (2000)

[23] Rayson, P., Leech, G., Hodges, M.: Social differentiation in the use of English vocabulary: some analyses of the conversational component of the British National Corpus. Intl. Journal of Corpus Linguistics 2(1), 133–152 (1997)

[24] Hofland, K., Johansson, S.: Word frequencies in British and American English, NCCH, Bergen, Norway (1982)

[25] Rayson, P.: Matrix: A statistical method and software tool for linguistic analysis through corpus comparison, Ph.D. thesis, Lancaster University (2003)

[26] Holmes, D.I.: Authorship attribution, Computers and the humanities 28(2), 87–106 (1994)

[27] Juola, P., Sofko, J., Brennan, P.: A prototype for authorship attribution studies. Literary and Linguistic Computing 21, 169–178 (2006)

[28] Wmatrix (April 2008), http://ucrel.lancs.ac.uk/wmatrix/

[29] SpectorSoft 'Spector Pro' (April 2008), http://www.spectorsoft.com

[30] Protecting Each Other, Crisp (April 2008),
     http://www.protectingeachother.com/

[31] SpyTech 'Spy Agent' (April 2008), http://www.spytech-web.com

[32] Rayson, P., Garside, R.: Comparing corpora using frequency profiling. In: Proceedings of the workshop on Comparing Corpora, held in conjunction with ACL 2000, Hong Kong, October 1-8, pp. 1–6 (2000)

[33] Sawyer, P., Rayson, P., Cosh, K.: Shallow Knowledge as an Aid to Deep Understand-ing in Early Phase Requirements Engineering. IEEE Transactions on Software Engineer-ing 31(11), 969–981 (2005)

[34] Face Book Press Information,
     http://www.facebook.com/press/info.php?statistics

[35] Peng, H.: A Data Mining Approach Based on Grey Prediction Model in Web Environment. Semantics, Knowledge and Grid, 76 (2006)

An Intelligent System for Semantic Information Retrieval Information from Textual Web Documents

Mukundan Karthik[1], Mariappan Marikkannan[2], and Arputharaj Kannan[3]

[1,3] Dept. of Computer Science, College of Engineering, Anna University, Chennai – 600025
[2] Dept. of Computer Science and Engineering, I.R.T.T, Erode – 638 316
karthikstars@gmail.com, mmk.irtt@gmail.com, kannan@annauniv.edu

Abstract. Text data, which are represented as free text in World Wide Web (WWW), are inherently unstructured and hence it becomes difficult to directly process the text data by computer programs. There has been great interest in text mining techniques recently for helping users to quickly gain knowledge from the Web. Text mining technologies usually involve tasks such as text refining which transforms free text into an intermediate representation form which is machine-processable and knowledge distillation which deduces patterns or knowledge from the intermediate form. These text representation methodologies consider documents as bags of words and ignore the meanings and ideas their authors want to convey. As terms are treated as individual items in such simplistic representations, terms lose their semantic relations and texts lose their original meanings. In this paper, we propose a system that overcomes the limitations of the existing technologies to retrieve the information from the knowledge discovered through data mining based on the detailed meanings of the text. For this, we propose a Knowledge representation technique, which uses Resources Description Framework (RDF) metadata to represent the semantic relations, which are extracted from textual web document using natural language processing techniques. The main objective of the creation of RDF metadata in this system is to have flexibility for easy retrieval of the semantic information effectively. We also propose an effective SEMantic INformation RETrieval algorithm called SEMINRET algorithm. The experimental results obtained from this system show that the computations of Precision and Recall in RDF databases are highly accurate when compared to XML databases. Moreover, it is observed from our experiments that the document retrieval from the RDF database is more efficient than the document retrieval using XML databases.

Keywords: Semantic relations, SEMINRET algorithm, Text mining, Resources Description Framework (RDF), Information Extraction (IE), Part-Of-Speech (POS) tag Intelligent Information Retrieval.

1 Introduction

Information Extraction (IE) [2] is a process, which takes unseen texts as input and produces unambiguous data in fixed-format as output. It involves processing text to identify selected information, such as particular named entity or relations among them from text documents. Named entities include people, organizations, locations and so

S.N. Srihari and K. Franke (Eds.): IWCF 2008, LNCS 5158, pp. 135–146, 2008.

on, while relations typically include physical relations (located, near, part-whole, etc.), personal or social relations (business, family, etc.), and membership (employ-staff, member-of-group, etc.).

With the explosive growth of the World Wide Web, we face an increasing amount of information resources, of which most are represented in free text [1]. Huge volumes of data can be accumulated beyond databases and data warehouses. Typical examples include the data streams, where data flow in and out like streams, as in applications like video surveillance, telecommunication, and sensor networks. The effective and efficient analysis of data in such different forms becomes a challenging task. The abundance of data, coupled with the need for powerful data analysis tools, has been described as a data rich but information poor situation.

Conventional systems use XML for representing web data. However, XML based representation has a number of limitations. Therefore it is necessary to use a better representation technique than the conventional XML format. Resources Description Framework (RDF)[1], proposed by the World Wide Web Consortium (W3C), is a language specification for modeling machine-processable and human-readable semantic metadata to describe Web resources on the Semantic Web. The basic element of RDF is RDF statements, which are triplets in the form of <subject, predicate, object>. An RDF statement can express that there is a relation (represented by the predicate) between the subject and the object.

Hence in this paper, we propose the use of RDF format for effective representation and manipulation of text data. The extracted semantic relations have been obtained in RDF Meta data form, since the RDF model is the canonicalization of a (directed) graph, and thereby it has all the advantages (and generality) of structuring information using graphs. Moreover, the RDF Meta data has the features, which includes common exchange syntax, handling partial information and consistency. From the RDF metadata, we can retrieve the semantic information effectively rather than the XML databases. The main difference between RDF and XML is that RDF has a higher abstraction level and RDF is an application of XML to represent metadata. Moreover, XML has a ordered tree like structure against the graph structure of RDF. At semantic level, in XML the information about the data is part of the data; in contrast RDF expresses explicitly the information about the data using relations between entities. An important advantage of RDF is its extensibility in both schema and instance level. In this paper, first we present a systematic approach and then explain how we have implemented an SEMINRET algorithm for semantic information retrieval from textual content. The rest of the paper is organized as follows. Section 2 explains the related works of the extraction of semantic relation from text documents. Section 3 depicts the proposed system architecture and their functionalities. Section 4 explains the implementation details of the semantic information retrieval algorithm. Section 5 describes the results discussion and comparative analysis for the proposed system. Finally, Section 6 concludes this paper with future research.

2 Related Works

There are many works in the literature that deal with Information Retrieval and Extraction. Among them, Zhou et al [3] introduced a technique for information retrieval using diverse lexical, syntactic and semantic knowledge for feature-based relation extraction using SVM. The feature subsystem covers word, entity type, overlap, base phrase chunking, dependency tree and parse tree, together with relation-specific semantic resources, such as country name list, personal relative trigger word list. Their results show that the feature- based approach outperforms tree kernel-based approaches, achieving 55.5% F- measure in relation detection and classification on the ACE2003 training data.

Ting Wang et. al [4] investigated a SVM-based classification for relation extraction and explored a diverse set of NLP features including Part-Of-Speech (POS) tag, entity subtype and class features, entity mention role features and general semantic features which all contribute to performance improvements. They also investigated the impact of different types of feature and different relation levels. N.Guarino, C.Masolo and G.Vetere [5] have proposed that representing the semantic relations by full conceptual graph standard, which is complex for large-scale applications. However, only simplified conceptual graphs are used in many existing practices in order to reduce the complexity.

A novel technique presented in [1] takes the mined results as input and produce a set of actions that can be applied to transform customers from undesirable classes to desirable ones. For decision trees, they have considered two broad cases. The first case corresponds to unlimited resources, and the second case corresponds to the limited resource-constraint situations. In both cases, their aim is to maximize the expected net profit of all the customers. Moreover, they have found a greedy heuristic algorithm to solve both problems efficiently and presented an ensemble-based decision-tree algorithm that use a collection of decision trees, rather than a single tree, to generate the actions. They show that the resultant action set is indeed more robust with respect to training data changes.

Sesame [16] acts as a backend for information storage and retrieval. WordNet is used for information regarding English language terms, while TAP [17] lends the system an understanding of common entities and helps determine their relation to the information stored in WordNet. Sesame is an open source semantic database that contains support for both schema inference and querying. It was chosen as the database because of its flexibility in terms of store and access methods in addition to its speed. TAP is quite similar to WordNet but with a differing focus, whereas WordNet contains dictionary terms. Moreover, TAP contains numerous real world entities ranging from people and places to even some of the more recent electronic devices. Combining these two knowledge bases allows the system to have an inbuilt understanding of an entity.

Current web information retrieval is based on the keyword search that has many well-known limitations. The reason is that most of web documents are in human-oriented formats (HTML, PDF, RTF etc.), which are suitable for the presentation, but machines cannot understand the meaning of published information. The semantic web technologies are capable of describing a meaning of web page information in a machine-understandable way. The semantic web [18] is meant to be an extension of the current web, in which existing web documents are annotated by machine-understandable metadata. T. Berners-Lee, J. Hendler, and O. Lassila, [6] described

Resources Description Framework (RDF) as a language specification for modeling machine-processable and human-readable semantic metadata to describe Web resources on the Semantic Web. An RDF statement can express that there is a relation (represented by the predicate) between the subject and the object. In [7], Berners-Lee further illustrated that RDF can be interworkable with conceptual graphs, [8],[9], which serve as an intermediate language for translating natural languages into computer-oriented formalisms. Resource Description Framework (RDF) [10] is a specification proposed by the World Wide Web Consortium (W3C) for describing and interchanging semantic metadata. The basic element of RDF is RDF statements, each consisting of a subject, a predicate, and an object. RDF is mainly a language specification addressing syntactical aspects. In this research work, RDF Schema [11] (RDFS) is further proposed to define RDF vocabularies for constructing RDF statements. In this work, we use RDF and RDFS to encode concepts and semantic relations which are extracted from textual Web content. Using NLP to derive Meta language features has been shown to benefit IE results [12]. Features, which have been used for relation extractions, include word, entity type, mention level, overlap, chunks, syntactic parse trees, and dependency relations [13] [14] [15].

In this paper, we propose an intelligent system for effective semantic information retrieval from the text documents, which is different from the existing works in many ways. In preprocessing of Textual web content that includes removal of Tags, Tokenization of the text and Tagging, we use the standard Universal Networking Language (UNL) tags for providing consistency with other systems. After the texts are tagged, they are parsed to generate the sentence grammar trees. In this step, for each sentence Noun and Verb Phrases are identified using Domain Lexicons. These are then used for extracting semantic relations. The sentence grammar trees are now parsed at each and every level to find terms and relation. The Extracted terms are now encoded in the form of metadata (RDF) which is not used in most of the systems. Finally the information is retrieved with the help of an intelligent system from the RDF using a special algorithm called SEMINRET (SEMantic INformation RETrieval) and compares the effectiveness of retrieval in RDF from other databases such as XML.

3   System Architecture

The architecture of the system developed in this work is shown in Figure 1. This system consists of four modules namely Text preprocessing, Semantic Relation Extraction, RDF metadata and Intelligent Information Retrieval module.

Text preprocessing: In this module, the query text contents are separated into Tokens for further processing which includes Tokenization, Tag removal and Tagging. POS Tagger is used to scan these tokens and assigns POS Tags. Also this POS Tagging is used to classify the words into the proper parts speech such as noun, verb, adjective, adverb, conjunction, determiner, pronoun and preposition. The POS tag and its types used from UNL standard are shown in Table – 1.

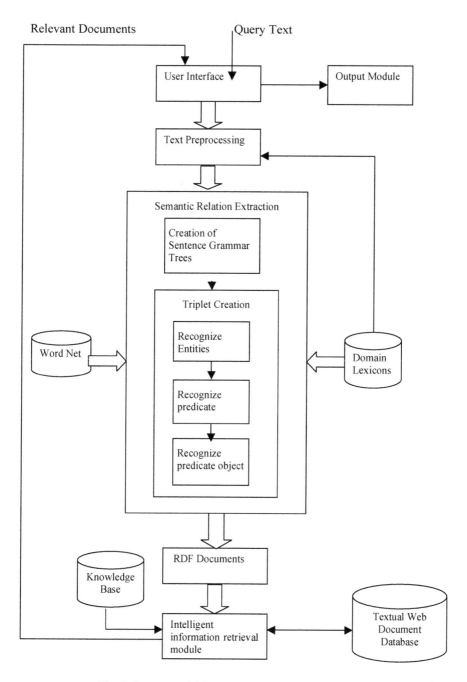

Fig. 1. System model for Semantic Information Retrieval

Table 1. Various Tags and its types

Type	Tag	Type	Tag
Noun	/NN	Conjunction	/CC
Plural Noun	/NNS	Determiner	/DT
Proper Noun	/NNP	Preposition	/IN
Verb	/VB	Adjective	/JJ
Verb (Present)	/VBP	Personal Pronoun	/PRP
Verb (Past	/VBD	Adverb	/RB
Verb (3rd singular)	/VBZ	Cardinal Number	/CD
Verb (Past Participle)	/VBD	Foreign Word	/FW
Verb (Present Participle)	/VBG	Symbol	/SYM

Extracting Semantic Relations: This module involves extracting terms and relations by parsing each and every level of the sentence grammar trees. This module is implemented using two main steps namely the creation of Sentence Grammar Trees and Triplets. The sentence grammar trees are used to find the entities and their relationship between them. The Triplet creation enables to recognize the entities, predicates and objects. The output of the Semantic Relation Extraction module is obtained in the form of RDF metadata which is fed into the Intelligent Information Retrieval Module.

Intelligent Information Retrieval Module: This module gets the RDF metadata as the input , applies the rules available in the Knowledge Base for effective storage and retrieval of Textual Web Documents.

4   Implementation Details

The given raw query text which are given from the user has been preprocessed in terms of Tokenization, Tag removal and Tagging. Then the extraction of semantic relations has been carried out through the following two steps:

I Creation of Sentence Grammar
This process converts the text into a tree structure that begins at a root node and progresses downward to the leaf nodes based on phrasal dependence. This method uses tagged text, which is broken down into the following three levels. They represent the grammatical usage from most inclusive to least inclusive:

1. *Clause Level*: This allows for full tagging of a coherent clause, the most common of
which is a single declarative clause represented as S.

Example: S → NP VP

2. *Phrase Level*: This enables phrase recognition within a clause. The most common of these as well as the most important components in this work are noun and verb phrases represented as NP and VP respectively.

Example: NP → Det PRO
VP → V NP

3. *Word Level*: This level simply provides the part of speech of each word in the sentence. For example: proper noun, noun, plural noun, etc. Relation Extraction relies, primarily, on phrase level tags. These tags provide us with indications on precisely how a given natural language sentence can be reformed into the subject -predicate - object triplet associated with the Semantic Web movement.

II Triplet creation: This module enables to recognize the entities, predicates and objects. Recognize entities from Sentence Grammar Trees.

In this step, Noun phrases are extracted from the sentence tree and are scanned for entities based on part-of-speech recognition. When a noun phrase is present in the phrasal sentence tree, it is understood to have the possibility of containing a single entity. Abstract concepts and real world objects are noted as entities.

Once the noun phrase has been extracted from the tree, it is parsed and nouns in both their singular and plural forms are combined to form a single entity. Specifically this means words fully contained within a noun phrase and tagged with NN - noun, NNS - plural noun, NNP – proper noun, NNPS - plural proper noun, or JJ - adjective. With the noun phrase extracted, it is parsed and nouns in both their singular and plural forms are combined to form a single entity. Now by using Word Net, each entity is assigned an URL.

Recognizing predicates.

In a manner similar to the extraction of noun phrases in the previous step, verb phrases are likewise generated from the tree. However, unlike noun phrases, verb phrases do not contain a single entity. In fact, all verb phrases, unless part of a sentence fragment, contain an underlying verb phrase and noun phrase, present in that order. These two included phrases are utilized to form the initial versions of the triplet predicate – object association.

The beginning verb phrase is recognized and parsed, words tagged with VB - verb, VBD - verb past tense, VBG - verb gerund, VBN - verb past participle, VBP - verb non-third person singular present and VBZ - verb third person singular present. At this stage, prepositions such as TO - to, determiners and adjectives are added to the predicate. The presence of words which are not tagged as verbs within the predicate such as prepositions, determiners, and adjectives are also given due importance in this step because they can alter the meaning of the sentence greatly.

Recognize Predicate Object.

Prepositional phrases are used further to alter the predicate - object pairing created in the previous step. This often involves combining the previous predicate - object linking with the prepositional phrase to create a new predicate. A new object is then formed from what remains of the verb phrase.

Now by generating the entities, predicates and objects the triplets are formed for statements, which can be converted into the form of RDF meta data. Finally the semantic information are retrieved from the RDF databases, by using the algorithm called SEMINRET (SEMantic INformation RETrieval) which is proposed and implemented in this paper.

The following steps are involved in the proposed **SEM**antic **IN**formation **RET**rieval (**SEMINRET**) algorithm for retrieving and extracting the semantic information from textual web document:

SEMINRET Algorithm:
1. Get the raw text from user.
2. Perform the text preprocessing using the following steps.
    a. Carry out the separation of text contents as tokens.
    b. Perform the tagging of each sentence.
    c. Using POS tagger, scan the tokens and assign the POS tag.
    d. Parse the tagged sentence to generate sentence grammar tree.
3. Find the terms and relations from the sentence grammar tree.
4. Extract the semantic relation with the coordination of Word net and Domain lexicon.
5. Obtain the semantic relation in the RDF metadata format.
6. Retrieve the intelligent semantic information from the RDF metadata using the rules present in the Knowledge Base.

The above algorithm has been explained below with a simple sentence and the outputs of each step are discussed as follows:

1. Raw query text:
    "India defeated Pakistan in World Cup Quarter final"
2. After the completion of text preprocessing, the generated sentence grammar tree for our example:

```
(ROOT
  (S
    (NP (NNP India))
    (VP (VBD defeated)
      (S
        (NP
          (NP (NNP Pakistan))
          (PP (IN   in)
            (NP (DT the) (NNP World) (NNP cup)
(NNP Quarter))))
        (ADJP (JJ final))))
    (.   .)))
```

3. The extracted semantic relation in the RDF triplet form is as follows:

India <defeated> Pakistan
India <defeated Pakistan in>World cup Quarter
India <final> World Cup Quarter

Finally the RDF Meta data is obtained, which is highly reliable and efficient than the existing methods. By using SEMINRET algorithm, we have obtained the effective result for semantic information retrieval.

5   Results and Discussions

The Web database used in this work has been constructed by collecting 10000 web pages from the World Wide Web (WWW). A query is given with different forms and the numbers of documents retrieved are given with rounded values near to thousands. The Table 2 shows the comparison between the number of relevant documents retrieved from the RDF and XML databases for the same query.

Table 2. Relevant Document Retrieval

Number of Documents Retrieved	Number of Relevant Documents Retrieved	
	In XML	In RDF
1000	450	546
2000	1078	1789
3000	2098	2896
4000	3098	3678
5000	3198	4983
6000	3209	5987

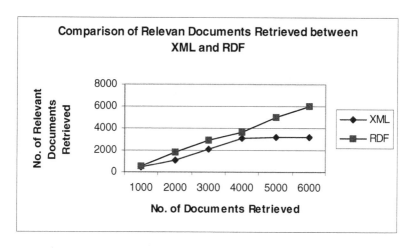

Fig. 2. Comparison graph between XML and RDF documents

Figure 2 shows that the comparison between the number of relevant documents retrieved between XML and RDF. From this graph, it is observed that the document retrieval from the RDF database is more efficient than the document retrieval using XML database.

Equations (1) and (2) show the method of computing precision and recall.

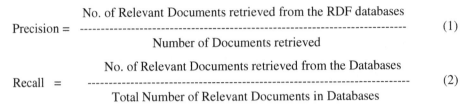

$$\text{Precision} = \frac{\text{No. of Relevant Documents retrieved from the RDF databases}}{\text{Number of Documents retrieved}} \quad (1)$$

$$\text{Recall} = \frac{\text{No. of Relevant Documents retrieved from the Databases}}{\text{Total Number of Relevant Documents in Databases}} \quad (2)$$

Figure 3 and Figure 4 illustrate graphically through line chart, the computed value of precision and recall for the set of number of documents. From the computed value of precision and Recall, the average accuracy of documents retrieval has been

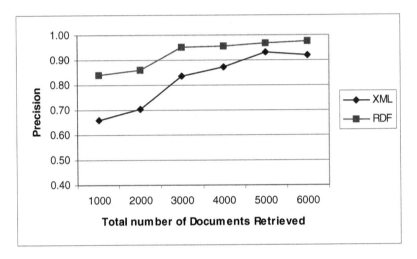

Fig. 3. Precision comparison

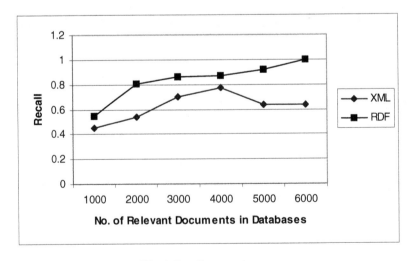

Fig. 4. Recall comparison

obtained and compared between XML and RDF databases. The computed values of Precision and Recall in RDF databases are highly accurate when compared to XML databases as shown in Figure 3 and Figure 4.

6   Conclusions and Future Enhancements

In this paper, we proposed a system for retrieving semantic information from the textual web document, which uses an intelligent retrieval algorithm for effective information retrieval from RDF databases. In this work, we computed the value of precision and recall for the set of number of documents. From the computed value of precision and Recall, the average accuracy of documents retrieval has been obtained and compared between XML and RDF databases. The experimental results show that the values of Precision and Recall in RDF databases are higher when compared to XML databases. Moreover, it has been known that the document retrieval from the RDF database is 20% more efficient than the document retrieval using XML databases.

In the future work, it is possible to apply the data mining algorithms to find the patterns from the RDF metadata for effective analysis. Moreover, this work has been tested with limited data sets. Better accuracy can be obtained by testing this system with a large data set.

References

1. Jiang, T., Tan, A.-H., Senior Member, IEEE, Wang, K.: Mining Generalized Asso-ciations of Semantic Relations from Textual Web Content. IEEE Transactions on Knowledge and Data Engineering 19(2), 164–179 (2007)
2. Appelt, D.: An Introduction to Information Extraction. Artificial Intelligence Communications 12(3), 161–172 (1999)
3. Zhou, G., Su, J., Zhang, J., Zhang, M.: Combining Various Knowledge in Relation Extraction. In: Proceedings of the 43rd Annual Meeting of the Association for Computational Linguistics, pp. 427–434 (2005)
4. Wang, T., Li, Y., Bontcheva, K., Cunningham, H., Wang, J.: Automatic Extraction of Hierarchical Relations from Text. In: Sure, Y., Domingue, J. (eds.) ESWC 2006. LNCS, vol. 4011, pp. 215–229. Springer, Heidelberg (2006)
5. Guarino, N., Masolo, C., Vetere, G.: Ontoseek: Content-Based Access to the Web. IEEE Intelligent Systems 14(3), 70–80 (1999)
6. Berners-Lee, T., Hendler, J., Lassila, O.: Semantic Web 284(5), 35–43 (2001)
7. Berners-Lee, T.: Conceptual Graphs and Semantic Web—Reflections on Web Architecture (2001), http://www.w3.org/DesignIssues/CG.html
8. Sowa, J.F.: Conceptual Structures: Information Processing in Mind and Machine. Addison-Wesley Longman, Amsterdam (1984)
9. Sowa, J.F.: Conceptual Graphs: Draft Proposed American National Standard. In: Proceeding of International Conference on Computational Science, pp. 1–65 (1999)
10. W3C, W3c RDF Specification (2005), http://www.w3.org/RDF/
11. W3C, W3c RDF Schema Specification (2005), http://www.w3.org/TR/rdf-schema/

12. Li, Y., Bontcheva, K., Cunningham, H.: SVM Based Learning System For Information Extraction. In: Mauri, G., Păun, G., Jesús Pérez-Jímenez, M., Rozenberg, G., Salomaa, A. (eds.) WMC 2004. LNCS, vol. 3365, pp. 319–339. Springer, Heidelberg (2005)
13. Freitag, D., McCallum, A.: Information extraction with HMM structures learned by stochastic optimization. In: Proceedings of the 7th Conference on Artificial Intelligence (AAAI 2000) and of the12th Conference on Innovative Applications of Artificial Intelligence (IAAI 2000), pp. 584–589. AAAI Press, Menlo Park (2000)
14. Zelenko, D., Aone, C., Richardella, A.: Kernel methods for relation extraction. Journal of Machine Learning Research, 1083–1106 (2003)
15. Zhou, G., Su, J., Zhang, J., Zhang, M.: Combining Various Knowledge in Relation Extraction. In: Proceedings of the 43rd Annual Meeting of the Association for Computational Linguistics (2005)
16. Broekstra, J., Kampan, A., van Harmelen, F.: Sesame: A generic architecture for storing and querying RDF and RDF Schema. In: International Semantic Web Conference, pp. 54–68 (2002)
17. Guha, R., McCool, R.: Tap: A Semantic Web Platform. Computer Networks 42(5), 557–577 (2003)
18. Berners-Lee, T., Hendler, J., Lassila, O.: The Semantic Web. Scientific American (2001)

Support of Interviewing Techniques in Physical Access Control Systems

Svetlana N. Yanushkevich[1], Oleg Boulanov[1],
Adrian Stoica[2], and Vlad P. Shmerko[1]

[1] University of Calgary, Biometric Technology Laboratory, Canada
[2] Humanoid Robotics Laboratory, California Institute of Technology,
Jet Propulsion Laboratory, NASA, Pasadena, CA, U.S.A.

Abstract. This paper investigates the enhancing of the decision-making support in the physical access control systems through deployment of the interviewing techniques widely used in forensics. While most of the physiological and behavioral information contents of the interviewee's responses can only be detected by the well-trained interviewer and supported by special devices such as lie detector, this approach is not feasible in the application such as physical access control. An alternative is the surveillance devices and biometric sensors for non-invasive registration of biometric data. In this paper, we focus on extracting the physiological and behavioral information from the video and infrared facial images. Our experiments show that the interviewing and interrogation techniques armored using the biometric assistant devices, can be employed to support the user's decision-making.

1   Introduction

This paper presents a summary of the application of the forensic techniques, such as interviewing, in the developing of the next generation of physical access security systems (PASSes). The main feature of the PASS is the efficient support of security personnel enhanced with the situational awareness paradigm and intelligent tools. The rationale for our interest in using interviewing techniques in the PASS is as follows. The access authorization process is characterized by *insufficiency* of information. The result of a customer's identification is a decision under uncertainty made by the user. For example, it is impossible to recognize, through visual observation of an individual, any changes due to plastic surgery; in other words, to determine whether the individual is trying to pass him/herself off as someone else. This situation is characterized by *insufficiency* of information. Uncertainty (incompleteness, imprecision, contradiction, vagueness, unreliability) is understood in the sense that the available information allows for several possible interpretations, and it is not entirely certain which is the correct one. The user must make a *decision under uncertainty*, that is, select an alternative before any complete knowledge is obtained. The use of access control systems rely on document check and biometric technologies, but they also utilize a dialogue of a user and a customer to receive more information for the customer

S.N. Srihari and K. Franke (Eds.): IWCF 2008, LNCS 5158, pp. 147–158, 2008.
© Springer-Verlag Berlin Heidelberg 2008

identification. The dialogue techniques refer to the forensic techniques, such as interviewing and interrogation. The verbal responses of the customer are used to make a decision. However, most physiological and behavioral information of these responses can only be detected by special devices such as lie detector, or partially detected by the well-trained user. Obviously, lie detector cannot be used in the application such as physical access control, while the surveillance devices and sensors for registration of data at distance can be utilized.

Several recent research initiatives, including the Defense Advanced Research Projects Agency (DARPA) research program, HumanID, promote the *early-detection* information, that is, the detection, recognition and identification of humans at a distance [1]. This information can help the user in interviewing of customer and improve the reliability of decision making. Specifically, in our design approach, we focus on deriving information from infrared and video images, for further utilization of this knowledge in the dialog support tools.

The main goal of facial analysis in infrared band implemented in the PASS is detection physiological parameters such as temperature, blood flow rate and pressure, as well as physiological changes caused by alcohol and substances. In particular, detection of a flu at distance can be supported. In addition to the detected temperature, the decision making tools uses such information as flight direction, season of the year, epidemiologic and other relevant data. Another useful application of the infrared image analysis is detection of artificial accessories such as artificial hair and plastic surgery. This data provides valuable information to support interviewing the customer in the PASS. In particular, this paper contributes to the study of the following problem: *How the interviewing techniques used in the systems like PASS can be improved using data from video and infrared images*. In our study of this problem, we use various software and hardware tools including cameras in visible and infrared bands, biometric devices, image processing tools and Bayesian networks.

2   Biometrics and Dialogue Support

The structure of the proposed PASS is shown in Fig. 1. The system consists of sensors such as *cameras* in the visible and infrared bands, *processors*, a *knowledge domain converter*, a *dialogue support* device, and a *personal file* generating module. Three-level surveillance is used in the system.

These levels correspond to three intervals of the time of service, T: T_1 (pre-screening phase of service, or waiting), T_2 (collection information during an individual's movement from the pre-screened position to the officer's desk) and T_3 (screening phase involving the identification, document-check and authorization). The first time interval, T_1, is suitable for obtaining early warning information using surveillance. Note that in terms of physical space, the distance between the pre-screened and screened areas can be used to obtain extra information using, for example, gait biometrics.

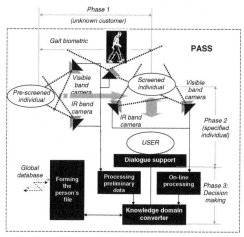

The structure of the proposed
PASS

▶ Cameras *in the visible and infrared bands*
▶ Processors *of preliminary information and
online data*
▶ Knowledge domain converter
▶ Dialogue support *device to support conversation based on the preliminary information obtained, and a module generating the
personal file.*
▶ *Three-level surveillance: the line (pre-
screening); the walk between pre-screened
and screened points, and the authorization
process at the user's desk (screening)*

Fig. 1. The PASS is a semi-automatic, application-specific distributed computer system
that aims to support the user's job in access authorization

The basic concept of the PASS is the collaboration of the user, the customer,
and the machine. This is a *dialogue-based interaction*. Based on the premise that
the user has priority in the decision making at the highest level of the system
hierarchy, the role of the machine is defined as assistance, or support of the user.
Three possible support strategies are introduced in Table 1. In our prototyping,
the last strategy is implemented; that is, the customer response is used for the
question generation.

Table 1. Support for cooperation in the PASS between user, customer, and machine

Cooperation	Specification	Representation
Supporting question-naire	The questionnaire technique of the user U is supported by machine M over weak co-operation with customer C and user U	
Supporting answering	The answering technique of customer C is supported by machine M over weak coop-eration with user U and customer C	
Supporting dialogue	Questionnaire and answering techniques of user U and customer C are supported by machine M over weak cooperation with user U and customer C	

3 Scenarios of Decision-Making Support

The possible scenarios are divided into three groups: regular, non-standard, and extreme. Let us consider an example of a scenario, in which the system generates the following data about the screened person (Figure 2).

Protocol for person #45 under pre-screening	Protocol for person #45 under screening
Time: 12.00.00: Warning: level 04 Specification: Drug or alcohol intoxication, level 03 Possible action: 1. Database inquiry 2. Clarify in the dialogue	Time: 12.10.20: Warning: level 04 Specification: Drug or alcohol intoxication, level 03 Local database matching: positive Possible action: 1. Further inquiry using dialogue 2. Direct to the special inspection

Fig. 2. Protocol of pre-screening (left) and screening (right)

It follows from this protocol (Fig. 2, left) that the system evaluates the third level of warning using automatically measured drug or alcohol intoxication for the screened customer. A knowledge-based sub-system evaluates the risks and generates two possible solutions. The user can, in addition to the automated analysis, evaluate the images acquired in the visible and infrared spectra.

We distinguish scenarios that correspond to the results of the matching of the customer's data with data in local and global databases. This process is fully automated for some stationary conditions (Fig. 2, right). Note that data may

Protocol of the person #45 under screening	Protocol of the person #45 under screening (continuation)
Time 00.00.00: Warning, level 04 Specification: Drug or alcohol consumption, level 03 Local database matching: positive Proposed dialogue questions: Question 1: Do you need any medical assistance? Question 2: Any service problems during the flight? Question 3: Do you plan to rent a car? Question 4: Did you meet friends on board? Question 5: Did you consume wine or whisky aboard? Question 6: Do you have drugs in your luggage?	Level of trustworthiness of Question 1 is 02: Level of trustworthiness of Question 2 is 02: Level of trustworthiness of Question 3 is 03: Level of trustworthiness of Question 4 is 00: Level of trustworthiness of Question 5 is 03: Level of trustworthiness of Question 6 is 03: Possible action: 1.Direct to special inspection 2. Further inquiry using dialogue

Fig. 3. Protocol of the person during screening: the question generation (left) and their analysis (right) with corresponding level of trustworthiness

not always be available in the database - this is the worst case scenario, and knowledge-based support is vital in this case.

The example in Fig. 3, left, introduces a scenario based on the analysis of behavioral biometric data. The results of the automated analysis of behavioral information are presented to the user (Fig. 3, right). Let us assume that there are three classes of samples assigned to "Disability", "Alcohol intoxication", and "Normal". The following linguistic constructions can be generated by the system: Not enough data, but abnormality is detected, or Possible alcohol intoxication, or An individual with a disability.

4   Interviewing Techniques

Human-human interface is an interaction between humans using linguistic data representation. Carriers of linguistic data are voice and text in documents. Human-human interaction is supported by biometric devices. Knowledge generated by these devices include updated biometric parameters and additional information such as voice stress features and, perhaps some contact biometrics such as fingerprints.

The user must be able to communicate effectively with the customer in order to minimize uncertainty. Limited information will be obtained if the customer does not respond to inquiries or if his/her answers are not understood. We distinguish two types of uncertainty about the customer: the uncertainty that can be minimized by using customer responses, his/her documents, and information from databases; and the uncertainty of appearance (physiological and behavior) information such as specific features in the infrared facial image, gait, and voice. The last type of uncertainty can be minimized by specifically-oriented questionnaire techniques. These techniques have been used in criminology, in particular, for interviewing and interrogation. The output of each personal assistant is represented in semantic form. The objective of each semantic construction is the minimization of uncertainly, that is, (a) choosing an appropriate set of questions (expert support) from the database, (b) alleviating the errors and temporal faults of biometric devices, and (c) maximizing the correlation between various biometrics.

Deception can be defined as a semantic attack that is directed against the decision-making process. Technologies for preventing, detecting, and prosecuting semantic attacks are still in their infancy. Some techniques of forensic interviewing and interrogation formalism with elements of detecting the semantic attack, are useful in dialogue development [9]. In particular, in order to fulfil a particular role, friendly or stern linguistic constructions, such as, How are you, sir, after this long flight? and Please, explain why have you changed your appearance?

The user's task is to be alert for inconsistencies in customer's replies. A deceptive person generally finds that more and more lies are necessary as additional details are required, and the person either forgets what he/she has previously asserted or fabricates details that are not compatible with previous statements.

Questions requiring "Yes" or "No" answer are generated by biometric-based personal assistants for the purpose of summarizing and verifying information, but they are not used when seeking new information.

Different decision strategies can provide varying decisions because of their different philosophies for dealing with uncertainty. In the PASS prototyping, we use *Bayesian belief networks*.

5   Biometric Data Processing

As a rule, the most of emotional responses of the customer during the dialogue are hard to register by the user. This fact can significantly contribute in uncertainly of human decision-making. Our goal is to minimize this uncertainly by providing support to the user. In particular, some hidden behaviorial responses can be detected using infrared band camera. The data acquired using infrared camera, along with video recording, and sometimes audio recording, can reveal some physiological and behavioral features, similar to ones, estimated by the lie detector. However, in contrast to the latter, the surveillance-based techniques are non-invasive, which is dictated by the application-specific environment.

In this paper, we address the problem of extracting information helpful for early detection of the physiological and psycho-emotional patterns based on hyper-spectral skin texture images.

5.1   Hyperspectral Analysis

Consider, for example, the processing of a facial image in both the visual and infrared bands. Such processing is aimed at so-called *hyperspectral* analysis. Suppose that indirect computing results in the detection of a possible drug intoxication. This result should be considered as a hypothetical data characterized by uncertainty, because of the shortcomings of the algorithms, sensor imperfection, measurement errors, and other factors. This drawback can be alleviated using intelligent technology and, more importantly, utilized for performing a more reliable authorization of the individual. This is because the transformation of biometric data into a semantic representation is based on an assumption about the probabilistic nature of the captured information. The problem to be solved is to construct this semantic form using the appropriate target functions: to justify, to clarify, to alarm, etc. These intelligent evaluations of the above scenario are very useful for user, and can be classified as assistance in decision-making.

A quantitative analysis of human skin color and temperature distribution can reveal a wide range of physiological phenomena. For example, skin color and temperature can change due to drug or alcohol consumption, and physical exercises [6, 8].

Since the color of human skin can reveal distinct characteristics valuable for diagnostics, many authors reported theoretical and experimental studies of the optical properties of human skin, specifically, the mechanism of skin color formation [2, 12]. It has been demonstrated, in particular, that the dominant pigments in skin color formation are *melanin* and *hemoglobin*. Melanin and hemoglobin determine the color of the skin by selectively absorbing certain wavelengths of the incident light. The melanin has a dark brown color and predominates in the epidermal layer while the hemoglobin has a hue or purple color, depending on the oxygenation, and is found mainly in the dermal layer. The quantities of melanin and hemoglobin pigments in the human skin were experimentally determined in [10, 11].

Distance infrared analysis of chemical composition, such as ethanol (alcohol), acetaminophen (major ingredient of tylenol), and codeine, in body fluids is an area of particular interest in medical and other applications. However, the detection of the actions of chemicals in the infrared, such as drug applications to the skin, have not been studied enough. In the early stages of some chemical actions, capillaries near the skin surface become enlarged and hot. As the reaction increases, a warm expanding network of capillaries (the area affected) can be observed in the infrared [5]. We are working on the hypothesis that facial infrared images are affected by specific drugs and alcohol.

We propose the so-called behavior indicator, the tools that derive the quantitive information based on the infrared surveillance. Let R_1 and R_2 be the responses of the customer to the questions Q_1 and Q_2, respectively. Responses R_1 and R_2 can be numerically estimated using facial infrared images. The *behavior indicator* is defined as the difference

$$\text{Behavior indicator } \Delta = R_1 - R_2$$

This difference can be interpreted as follows. One of the questions (Q_1 or Q_2, depends on the sign of Δ) was more difficult then another; that is, to answer the "hard" question, it takes a customer more emotional or/and intellectual efforts.

This well-known effect is utilized in the interviewing and interrogation to detect emotional changes, in particular, in lie detectors [4]. It was experimentally justified in [3], that infrared facial images can be used as the additional or alternative valuable source for detection of emotional changes through interrogation.

The behavior indicator captures two behavioral characteristics: (a) the emotional state and (b) the amount of intellectual work. In some approaches, these two characteristics are not distinguished [3, 4]. However, we believer that distinguishing these two different psychological characteristics can be useful in minimization of risks of wrong decisions in polygraph-based forensic techniques and in the PASS. For example, detection of high intellectual efforts (without emotional changes) to answer the simple question can be used as indicator for the correction or change of interviewing or/and interrogation strategy.

6   Experimental Results

The devices gathered from the sensors and intelligent data processing for the situational awareness are called in our approach the *decision-making support assistants*. These assistants can be based on non-invasive metrics such as: temperature measurement, artificial accessory detection, estimation of drug and

Surveillance video
camera JAI CV-M9

Surveillance infrared
camera Miricle KC 307K

▶ *Two JAI CV-M9 CL* $3 \times 1/3''$ *progressive scan RGB color cameras with* 1034×779 *4.65* μm *effective square pixels for each CCD.*
▶ *A Thermoteknix MIRICLE 307K uncooled microbolometer infrared camera with a focal plane array of* 640×480 *pixel size and a dynamic range of 14 bits. The spectral band of the camera is* $7-14\mu m$ *and the standard frame rate is* $25-30$ *frames per second.*
▶ *A PC station with acquisition boards (Euresys GRABLINK Expert 2 for video cameras and Picolo Pro 2 for the thermal camera).*

Fig. 4. A setup of a pair of video and infrared cameras for surveillance

alcohol intoxication, and estimation of blood pressure and pulse. The basic design paradigm of these decision support assistances is the *discriminative* biometrics.

In our experiments, we focus on detection in facial infrared images the intellectual efforts needed for correct answer. We found that the behavior indicator can be computed for this purpose using analysis of video and infrared images.

Our experiments first concern with a video and infrared sensor-based assistants. A setup of the paired video and thermal cameras for acquisition of facial images in both the visual and infrared bands is shown in Fig. 4.

For skin analysis and synthesis, we adopted the model proposed in [6, 8] as a basis, and developed it further in order to meet its requirements in the context of an early detection support system. Specifically, we studied the spacial and

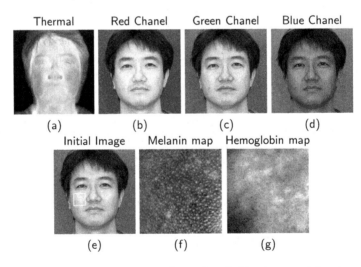

Thermal   Red Chanel   Green Chanel   Blue Chanel

(a)   (b)   (c)   (d)

Initial Image   Melanin map   Hemoglobin map

(e)   (f)   (g)

Fig. 5. Visual information contained in thermal and video images: (a) a thermal image and the (b) red, (c) green, and (d) blue components of a video image; Extraction of melanin and hemoglobin information by ICA: (e) video image with the selected region, (f) melanin, and (g) hemoglobin components for the selected region

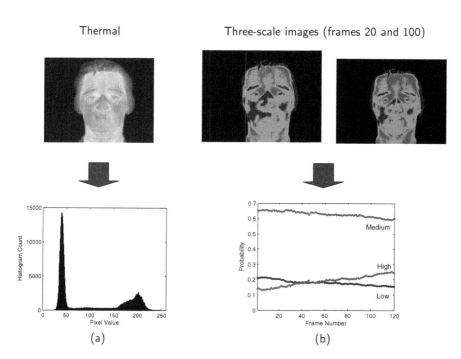

Fig. 6. Dynamics of thermal images due to a mental effort: (a) a thermal image and the histogram in which the region between 156 and 225 pixel values corresponds to the face region; (b) three-scale images (the images in which the pixels are distributed according to three temperature ranges: medium, high and low, indicated by different colors) corresponding to the 20th and 100th frames of the thermal images, and the graph of proportions (called here the probability) of the pixels from three temperature ranges

temporal structure of facial images and the correlation of the skin color with the skin temperature changes acquired by the thermal camera. As far as skin color formation is concerned, we adopted the model proposed in [6, 7, 8] and represent colors in the optical density domain by a color vector indirectly representing Red-Green-Blue (RGB) values. The color vectors represent the are intrinsic characteristics of the pigments and do not alter from one point to another; it is the changes of the quantities of hemoglobin and melanin that produce all the rich variations in skin color. The color space of the skin is two-dimensional and forms a surface in the 3D RGB color space. We can reduce the dimensionality of the problem by applying PCA prior to ICA. The texture maps representing the hemoglobin and melanin content of the facial skin are analyzed to find a correlation with the temperature distribution rendered from thermal images.

An example of visual information acquired by the cameras is shown in Fig. 5. Fig. 5(a) shows a thermal image acquired by the system and Figs. 5(b)–(d) demonstrate the RGB color components of a video image. The RGB components are used as input signals for ICA of the skin texture maps. Thus, Fig. 5(e)–(g) illustrate the melanin and hemoglobin maps extracted from a selected rectangular area.

This information is used in the currently developed approach to fusion of visual and infrared facial image information for evaluating the physiological and psycho-emotional state of a person.

Our study also concerned with studying dynamics of infrared images during interviewing. The interval of observation is used to record a thermal video, and then analyze frames taken using regular intervals. The simplest analysis involves count the number of pixels, correspondingly to the low, medium, and high temperature, taken as a proportion to the total number of facial image pixels. The first image in Fig. 6b is taken in the beginning of performing the calculation, and the second image is taken at the end. The proportion of the number pixels in each region to the total number of pixels (called probability) is changed during thermal video recording during calculation.

We simplified our experimental study because of the complexity and high cost of real-world experiments: instead of observing responses to questions, we asked the tested person to solve various mathematical calculations. Similarly to the questionnaire techniques, this required some intellectual effort. Based on this premise, we analyzed the dynamics of infrared images of people, participated in the study (Fig. 6). The primary conclusion is that facial images in infrared band can distinguish people in the relaxing state and people making calculation tasks.

7   Further Development: Training System

Prediction is one of the most important features of a training system, which models real world processes and estimates the effects using controlled events. In our approach, the design of an expensive training system is replaced by an inexpensive extension of the PASS, already deployed at the place of application [13, 14]. In this way, a long-term training is replaced by periodically repeated, short-term, intensive and computer-aided training.

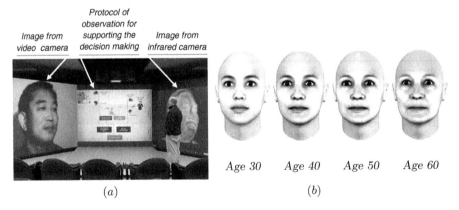

Fig. 7. Monitoring equipment for training personnel for the decision-making support system: (a) generated test tasks and (b) aging modeling (neutral facial expression) using the package FaceGen (Virtual Reality Room of the University of Calgary)

We utilized the *Silicon Graphics* facilities and monitoring equipment of the *Virtual Reality Room* of the University of Calgary (Fig. 7). Also, software tools for synthetic biometrics, such as the *FaceGen* package for face modeling, and the *Comnetix Life-Scan* station for facial image acquisition and identification. We developed an approach that alternates the known approaches with respect to several criteria, including cost-efficiency in personnel training. In particular, in training system, modeling is replaced by real-world conditions, and long term training is replaced by *periodically repeated short-term intensive* computer-aided training.

8   Conclusion

Our study concerns with developing an approach to support the user decision-making under uncertainty in the PASS. We focus on the scenario when dialogue techniques supported by biometric assistants are used to make the final decision. In some particular cases, the interviewing techniques are used in the existing PASSes. Our goal is to improve the application of these techniques using the PASS advantages such as usage of biometric data and the decision making support, while taking into account the PASS constraints, such as non-invasive data acquisition and the application-specific environment.

The biometric data can be acquired before and during interview as well as periodically updated during the interview. The dialogue itself can be supported using the well-developed in forensic and justice questionnaire techniques. The responses to the tasks of different complexity can differ, and this difference can be detected via analysis of facial images in visual as well as in infrared band.

Emotions contribute additionally to temperature distribution in the infrared facial image. The task for further study is to distinguish emotion response and response to the intellectual challenge task. Currently, we study the emotion contribution to the biometric data analysis results, as well as questionnaire techniques within the experimental conditions.

Acknowledgments

This Project is partially supported by the Natural Sciences and Engineering Research Council of Canada (NSERC), the Canadian Foundation for Innovations (CFI), the Government of the Province of Alberta, and the Alberta Informatics Circle of Excellence (iCore). A part of the project has been implemented as an initiative within the JPLs Humanoid Robotics Laboratory, NASA, USA.

References

[1] DAPRA: Total Information Awareness DAPRA's Research Program. Information and Security 10, 105–109 (2003)

[2] Dawson, J.B., Barker, D.J., Ellis, D.J., Cotterill, J.A., Grassam, E., Fisher, G.W., Feather, J.W.: A Theoretical and Experimental Study of Light Absorption and Scattering by in Vivo Skin. Phys. Med. Biol. 25(4), 695–709 (1980)

[3] Pavlidis, I., Levine, J.: Thermal Image Analysis for Polygraph Testing. IEEE Engineering in Medicine and Biology Magazine 21(6), 56–64 (2002)

[4] The Polygraph and Lie Detection. The National Academies Press, Washington, DC (2003)

[5] Prokoski, F.: History, Current Status and Future of Infrared Identification. In: Proc. IEEE Workshop on Computer Vision Beyond the Visible Spectrum: Methods and Application, pp. 5–14 (2000)

[6] Tsumura, N., Nakaguchi, T., Ojima, N., Takase, K., Okaguchi, S., Hori, K., Miyake, Y.: Image-Based Control of Skin Melanin Texture. Appl. Opt. 45, 6626–6633 (2006)

[7] Tsumura, N., Haneishi, H., Miyake, Y.: Independent-Component Analysis of Skin Color Image. J. Opt. Soc. Am. A 16, 2169–2176 (1999)

[8] Tsumura, N., Ojima, N., Sato, K., Shiraishi, M., Shimizu, H., Nabeshima, H., Akazaki, S., Hori, K., Miyake, Y.: Image-Based Skin Color and Texture Analysis/Synthesis by Extracting Hemoglobin and Melanin Information in the Skin. ACM Trans. Grap. 22(3), 770–779 (2003)

[9] Royal, R.F., Schutt, S.R.: The Gentle Art of Interviewing and Interrogation: A Professional Manual and Guide. Prentice-Hall, Englewood Cliffs (1976)

[10] Shimada, M., Yamada, Y., Itoh, M., Yatagai, T.: Melanin and Blood Concentration in Human Skin Studied by Multiple Regression Analysis: Experiments. Phys. Med. Biol. 46, 2385–2395 (2001)

[11] Shimada, M., Yamada, Y., Itoh, M., Yatagai, T.: Melanin and Blood Concentration in a Human Skin Model Studied by Multiple Regression Analysis: Assessment by Monte Carlo Simulation. Phys. Med. Biol. 46, 2397–2406 (2001)

[12] Van Gemert, M.J.C., Jacques, S.L., Sterenborg, H.J.C.M., Star, W.M.: Skin Optics. IEEE Trans. Biomedical Engineering 36(12), 1146–1154 (1989)

[13] Yanushkevich, S.N., Stoica, A., Shmerko, V.P.: Experience of Design and Prototyping of a Multi-Biometric Early Warning Physical Access Control Security System (PASS) and a Training System (T-PASS). In: Proc. 32nd Annual IEEE Industrial Electronics Society Conf., pp. 2347–2352 (2006)

[14] Yanushkevich, S.N., Stoica, A., Shmerko, V.P.: Fundamentals of Biometric-Based Training System Design. In: Yanushkevich, S.N., Wang, P., Srihari, S., Gavrilova, M., Nixon, M.S. (eds.) Image Pattern Recognition: Synthesis and Analysis in Biometrics. World Scientific, Singapore (2007)

Application of q-Gram Distance in Digital Forensic Search

Slobodan Petrović and Sverre Bakke

NISlab, Department of Computer Science and Media Technology,
Gjøvik University College, P.O. box 191, 2802 Gjøvik, Norway
{slobodanp, sverre.bakke}@hig.no

Abstract. In order to find evidence, digital forensic investigation often includes search procedures applied on large data sets. For such search procedures, appropriate fault tolerant distance measures are needed in order to detect evidence even if it has been previously distorted/partially erased from the search media. One of the appropriate fault-tolerant distance measures for this purpose is constrained edit distance, where the maximum numbers of consecutive insertions and deletions represent the constraints. However, the time complexity of its computation is too high. We propose a two-phase indexless search procedure for application in forensic evidence search that makes use of q-gram distance instead of the constrained edit distance. The q-gram distance is known to approximate well the *unconstrained* edit distance. We study how well q-gram distance approximates edit distance with special constraints needed in forensic search applications. We compare the performances of the search procedure with the two distances applied in it. Experimental results show that the procedure with the q-gram distance implemented achieves for some values of q almost the same accuracy as the one with the constrained edit distance, but the efficiency of the procedure that implements the q-gram distance is much better, for a much lower time complexity of computation of the q-gram distance.

Keywords: Forensic Search, q-gram distance, Constrained Edit distance.

1 Introduction

In order to find evidence, digital forensic investigation often includes search procedures applied on large data sets. In such scenarios, the capacities of today's electronic storage devices may represent a problem, since they often reach several terabytes [1] and are therefore a serious challenge even to the best known search algorithms.

The overall efficiency of current commercial digital forensic search tools is constrained by the employment of simple hashing and indexing algorithms [10]. On the other hand, content retrieval data mining techniques, which are either keyword similarity based or indexing based are studied intensively and are expected to find wide practical application.

It may be very useful to employ a fault tolerant distance measure in forensic content retrieval data mining (i.e. forensic search), since fault tolerant distance enables detection of the query (evidence string) even if it has been partially erased or distorted by inserting, substituting or deleting characters. One of the adequate fault tolerant distance

S.N. Srihari and K. Franke (Eds.): IWCF 2008, LNCS 5158, pp. 159–168, 2008.
© Springer-Verlag Berlin Heidelberg 2008

measures is the constrained edit distance, where the maximum numbers of consecutive insertions and/or deletions represent the constraints. However, the time complexity of the computation of the constrained edit distance is quadratic in the lengths of the involved strings. To improve the efficiency of the forensic search procedure, the constrained edit distance could be approximately computed by means of another distance measure whose computation is less complex. The q-gram distance may be useful for this purpose. It is known to approximate well the *unconstrained* edit distance [15]. However, there is no study that would evaluate quality of approximation of the constrained edit distance needed in forensic search by the q-gram distance.

In this paper, we apply the q-gram distance in the process of pre-selection of the forensic data search space fragments of the pre-determined length that are interesting for eventual matching of the particular evidence search pattern. A coarse pre-selection matching is performed first, by computing the q-gram distance between each forensic data fragment and the search query. The obtained q-gram distances are then sorted in the increasing order and a finer inspection is performed in the fragments starting from those at the minimum q-gram distance from the evidence search query.

We compare the accuracy of the search procedure in which the q-gram distance is applied with the accuracy of the same procedure in which the constrained edit distance is used over a sample 30 kB test data set. We experimentally show that for some values of the parameter q, the constrained edit distance is approximated very well with the q-gram distance.

1.1   Previous Work

Efficient algorithms with sublinear time complexities for exact string search were devised several decades ago [2,5]. These algorithms, however, cannot be used for fault-tolerant search. To solve this problem, edit distance based algorithms have been used. For example, in [4,9], procedures intended for general application are described that successfully detect distorted contiguous and non-contiguous subsequences in a given dictionary/database. Other general purpose approximate string matching techniques make use of (still unconstrained) weighted edit distance (see for example [6,12]). In approximate search in text databases in particular, the search techniques are usually based on the construction of an index data structure in advance that enables a faster response from the database at the time of submitting the query. For example, in [14] a special index data structure called V-tree is first constructed and then partitioned by means of a clustering technique in order to improve the efficiency of approximate search for words in text databases. Edit distance without constraints is also used in this system. In [13], the index is stored in the so-called monotonous bisector tree (MBT) and to further increase the efficiency of the search and pre-processing (i.e. computing the index) an approximate technique that estimates the unconstrained edit distance by means of q-gram distance is used. Many commercial forensic search tools also use indexing to speed-up the search (see for example [3]). These commercial programs allow certain level of tolerance in recognising the search query as well. The drawback of the indexing methods in general is the need for pre-processing (indexing) that is memory consuming and may be time-consuming, making them often unsuitable for constrained-time operations.

Mihov et al. [7,8] analyze possibilities of application of edit distance in lexical text correction. This problem is similar to approximate forensic search in a sense that we would like to find the strings in the search material whose edit distance from the search query is less than or equal to a limit given in advance. In order to overcome the problems related with too many alternatives offered by the procedure that employs unconstrained edit distance [7], in [8] specific constraints in the definition of edit distance are introduced. These constraints deal with permitted substitutions, not with the insertion/deletion constraints.

2   Mathematical Background

We first define the edit distance with the constraints as in [11]. Given arbitrary strings X and Y of finite lengths N and M, respectively, over a finite alphabet $\mathcal{A}$, we deal with the problem of transforming the string X to Y by means of the elementary edit operations of deletion, insertion, and substitution, under the following constraints:

C_1 The number of insertions belongs to a set T given in advance.
C_2 The maximum lengths of runs of deletions and insertions are F and G, respectively.
C_3 The edit sequence is ordered in a sense that every substitution is preceded by at most one run of deletions followed by at most one run of insertions.

The constrained edit distance $D(X, Y)$ is then defined as the minimum sum of elementary edit distances associated with the edit operations of deletion, insertion, and substitution needed to transform X to Y, subject to the assumed constraints.

In order to be applied in forensic search, the constraints in the definition above have to be removed *before the first substitution* and *after the last substitution*. That means that arbitrary numbers of deletions and insertions are allowed before the first substitution and after the last substitution.

We now proceed with the definition of the q-gram distance as the distance measure capable of approximating well the edit distance. Let $\mathcal{A}$ be a finite alphabet. Let $\mathcal{A}^*$ denote the set of all strings over $\mathcal{A}$ and let $\mathcal{A}^q$ denote the set of all the strings of length q over $\mathcal{A}$, $q = 1, 2, \ldots$ A q-gram is any string $v = v_1 v_2, \ldots v_q$ in $\mathcal{A}^q$.

Let $S = s_1 s_2, \ldots s_N$ be a string of length N in $\mathcal{A}^*$, and let v in $\mathcal{A}^q$ be a q-gram. If $s_i s_{i+1} \ldots s_{i+q-1} = v$ for some i, then S has an occurrence of v. Let $G(S)[v]$ denote the total number of occurrences of v in S. The q-gram profile of S is the vector $G_q(S) = (G(S)[v])$, $v \in \mathcal{A}^q$.

Let X and Y be strings of finite lengths N and M, respectively, over $\mathcal{A}^*$, and let $q > 0$ be an integer. The q-gram distance between X and Y is

$$D_q(X, Y) = \sum_{v \in \mathcal{A}^q} | G(X)[v] - G(Y)[v] | . \tag{1}$$

The intuition behind the approximation of the edit distance with the q-gram distance is that when two strings are within a small edit distance of each other, they share a large number of q-grams [15].

In [15], two methods for computing the q-gram distance are given. We present the method that involves integer encoding of q-grams, since it is often used in practice (see for example [13]). Let the alphabet be $\mathcal{A} = \{a_0, a_1, \ldots, a_{z-1}\}$, with $z = \mid \mathcal{A} \mid$. Any q-gram $v = v_1 v_2 \cdots v_q$ can be interpreted as an integer $f(v)$ in base-z notation:

$$f(v) = \sum_{i=1}^{q} \overline{v}_i z^{q-i}, \tag{2}$$

where $\overline{v}_i = j$ if $v_i = a_j$, $i = 1, 2, \ldots, q$, $j = 0, \ldots, z - 1$. $f(v)$ is a bijection, i.e. $f(v) = f(v')$ if and only if $v = v'$. $f(v)$ is called the fingerprint of v and can be used as the index of the q-gram profile array in the practical computation. Note that the length of the q-gram profile of any string over $\mathcal{A}$ is $\mid \mathcal{A} \mid^q$.

The q-gram profile of the string S of length N is then computed by getting the q-gram v at the i-th position of S, computing the corresponding index $f(v)$ by means of the expression (2) and finally incrementing the contents of the profile element at the position $f(v)$, for $i = 1, \ldots, N - q + 1$.

In [15], the following theorem was proved that determines the time and space complexities of the computation of the q-gram distance if integer encoding of the q-grams is used for that computation:

Theorem 1
The q-gram distance $D_q(X, Y)$ can be computed in time $O(N + M)$ and in space $O(\mid \mathcal{A} \mid^q + N + M)$.

The obvious advantage of the computation of the q-gram distance over the computation of the edit distance is in its linear time complexity in the sum of the lengths of the involved sequences, unlike the quadratic time complexity needed for the computation of the edit distance. The space complexity of the computation of the q-gram distance can be reduced by applying the standard hashing techniques, see [15].

3   The Forensic Search Algorithm

The forensic search algorithm consists of two phases. In the first, pre-selection phase, for each forensic data fragment of the pre-determined length N, the q-gram distance given by (1) between the fragment and the search string is computed. The q-gram distances obtained in this way are sorted in ascending order. In the second phase of the search process, a finer search is performed over the fragments starting from the top ranked one in the first phase. We concentrate on the first phase of the forensic search procedure, since the second phase that includes exhaustive search over the selected fragments is straightforward. The complete search algorithm used in the first phase is given below.

Algorithm 1
Input:

- N - the length of the fragment.
- The value of the parameter q.
- S - the search string.
- $\mathcal{D}$ - the forensic data set.

Output:

– The array **P** of ordered pairs (i, d), sorted by d in ascending order, where i is the ordinal number of the fragment in $\mathcal{D}$, and d is the corresponding q-gram distance between the search string and the fragment.

begin

    comment Initialization

    Partition $\mathcal{D}$ into $k = \left\lceil \dfrac{|\mathcal{D}|}{N} \right\rceil$ fragments ;

    comment $\mathcal{D}_i$ is the i-th fragment of $\mathcal{D}$, $i = 1, 2, \ldots, k$
    for $i \longleftarrow 1$ **until** k **do**
      begin
        comment Compute the q-gram distance
                between $\mathcal{D}_i$ and $\mathcal{S}$ (See Alg. 2)

        $d = Qdist(\mathcal{D}_i, \mathcal{S}, q)$;
        Store (i, d) in **P** ;
      end ;
    Sort the array **P** by d in ascending order ;

end.

The following algorithm is used to compute q-gram distance between two strings. It is based on the expressions (1) and (2).

Algorithm 2 (*Qdist* - get the q-gram distance)

Input:

– The sequences $\mathcal{P}$ and $\mathcal{S}$ of lengths N and M, respectively.
– The value of the parameter q.

Output:

– The q-gram distance d between $\mathcal{P}$ and $\mathcal{S}$.

begin

    comment Get the q-gram profiles first
    for $i \longleftarrow 1$ **until** $N - q + 1$ **do**
      begin
        $qGram = \text{GetQgram}(\mathcal{P}, i)$;
        index = QgramIndex($qGram$); (expression (2));

```
    profile𝒫[index]++;
end;
```

comment Do the same for $\mathcal{S}$ and thus get profile$\mathcal{S}$

comment Compute the q-gram distance

```
dist = 0;
for i ⟵ 0 until | 𝒜 |^q −1 do
    dist += | profile𝒫[i] - profile𝒮[i] |;
return dist;
```

end.

4   Experimental Work

Suppose we want to find all the occurrences of an evidence string in a large capacity computer memory unit. The fact that such a string has/has not been found would later be used in a court. Besides large capacity of the memory space to be browsed, an additional problem is that the time at the investigator's disposal is limited. In addition, we would like our search to be fault tolerant, i.e. the search string would be reported detected even if it were deliberately distorted up to the predefined extent.

In such a context, the application of the constrained edit distance based (or q-gram distance based, where the q-gram distance is used to approximate the constrained edit distance) two-phase search technique is possible. We first define the tolerance level Δ as the extent to which we accept the deviation of the edit/q-gram distance between the fragment and the search string from the minimum possible one. For any Δ, we accept the string if the constrained edit/q-gram distance between the fragment and the search string is $d \leq N - M + \Delta$, where N and M are the lengths of the fragment and the search string, respectively.

Example: Let $N = 1000$ and let the search string be "ROLEX", i.e. $M = 5$. Since $N > M$ and since we suppose that there are no reversals of the characters in the search string, we can assume that it is possible to transform the fragment string of length $N = 1000$ into the search string by using only deletions and substitutions of symbols. Let the maximum number of consecutive deletions in edit transformations be $F = 1$ and let $\Delta = 1$. Then we select a fragment for the second phase of the search algorithm if the constrained edit distance between the fragment and the search string is $d \leq 996$. The relevant portions of two possible acceptable edit sequences are given in Fig. 1.

In order to compare the accuracy of forensic search with the constrained edit distance and the q-gram distance, the following experiment was carried out: 100 different search strings, whose lengths were between 2 and 13, were searched for in a 30 kb text file, where for every fragment the constrained edit distance and the q-gram distance were computed between the fragment and the search string. The length N of the fragment varied between 100 and 1000. In the computation of the constrained edit distance, no

```
. . . . .ROLEX. . . . .
φφ. .φROLEXφ. .φφ    N=1000, M=5, d=995
```

```
. . . . .ROLEX. . . . .
φφ. .φROφEXφ. .φφ    N=1000, M=5, d=996
```

Fig. 1. Two acceptable edit sequences $F = 1$, $\Delta = 1$ (see text). ϕ is used to represent deletion.

insertions were used, since in this controlled environment there were no reversals among the edit transformations and the fragment was always longer than the search string.

For a threshold Δ given in advance, a fragment was accepted as a candidate for a more detailed search in the second phase of the search process if the constrained edit distance/q-gram distance between the fragment and the search string was less than or equal to $N - M + \Delta$, where N was the length of the fragment and M was the length of the search string. For each search string, the fragments that were accepted with the use of edit distance and the q-gram distance were labelled (1 if accepted, otherwise 0). Thus we got two binary vectors and at the end of the experiment the Hamming distance between these two vectors was computed. The length of these vectors varied significantly, since the number of fragments varied, depending on the length of the fragments, and since search strings shorter than q cannot be processed with the procedure that uses the q-gram distance. In general, the lengths of the binary vectors are equal to the number of used search strings multiplied by the number of fragments, which results in the variation of the lengths of the binary vectors between approx. 2700 and 29300. The parameter q in the experiment varied between 1 and 3, and the threshold Δ for both types of distances varied between 0 and 3. The maximum length of runs of deletions at the computation of the edit distance varied between 1 and 5. Thus, also bearing in mind the variation in the lengths of the fragments, 2400 combinations of parameters in the computation of the distances were possible.

Table 1 presents several best values of the Hamming distances obtained in the experiment as well as several worst values. It can be observed that the achieved accuracy obtained by use of the q-gram distance depends heavily on the choice of the parameters q and N. The best accuracy was obtained with relatively high values of q and relatively low values of N. Among the best results, Δ_q had lower values, but variation of Δ_q between 0 and 1 did not cause any change in accuracy. The same holds for the worst results, but the values of Δ_q here were higher and varied between 2 and 3. Approximation of the constrained edit distance by means of the q-gram distance gave search error rate less than 2% for 32 combinations of the distance computation parameters. This indicates that with high probability low error rates can be expected for general data as well.

It can be concluded from the experiment that it is possible to achieve very high accuracy with the use of the q-gram distance, for relatively high values of q and the values of N and Δ that are often used in practice. Since the alphabet used in general files (not necessarily text files) may be of relatively high cardinality, in order to use even higher values of q some hashing techniques in the computation of the q-gram

Table 1. Accuracy obtained by use of q-gram distance (len - length of the binary vector (see text))

No.	q	Δ_q	F	Δ_e	N	len	d_H	ε [%]
1	3	0	1	0	100	27542	99	0.359
2	3	1	1	0	100	27542	99	0.359
3	3	0	1	0	200	13724	99	0.721
4	3	1	1	0	200	13724	99	0.721
5	3	0	2	0	100	27542	201	0.730
6	3	1	2	0	100	27542	201	0.730
7	2	0	1	0	100	29300	259	0.884
8	2	1	1	0	100	29300	259	0.884
9	3	0	1	0	300	9118	92	1.009
10	3	1	1	0	300	9118	92	1.009
11	3	0	3	0	100	27542	316	1.147
12	3	1	3	0	100	27542	316	1.147
13	2	0	2	0	100	29300	359	1.225
14	2	1	2	0	100	29300	359	1.225
15	3	0	2	0	200	13724	185	1.348
16	3	1	2	0	200	13724	185	1.348
17	3	0	1	0	400	6862	93	1.355
18	3	1	1	0	400	6862	93	1.355
19	3	0	1	0	500	5452	81	1.486
20	3	1	1	0	500	5452	81	1.486
21	3	0	4	0	100	27542	424	1.539
22	3	1	4	0	100	27542	424	1.539
23	2	0	3	0	100	29300	462	1.577
24	2	1	3	0	100	29300	462	1.577
25	3	0	1	0	600	4512	77	1.707
26	3	1	1	0	600	4512	77	1.707
27	3	0	2	0	300	9118	165	1.810
28	3	1	2	0	300	9118	165	1.810
29	2	0	4	0	100	29300	558	1.904
30	2	1	4	0	100	29300	558	1.904
31	3	0	5	0	100	27542	543	1.972
32	3	1	5	0	100	27542	543	1.972
...	...	...	...	...	...	...	...	...
2391	1	2	1	0	600	4800	3250	67.708
2392	1	3	1	0	600	4800	3250	67.708
2393	1	2	1	0	700	4100	2825	68.902
2394	1	3	1	0	700	4100	2825	68.902
2395	1	2	2	0	900	3200	2212	69.125
2396	1	3	2	0	900	3200	2212	69.125
2397	1	2	1	0	1000	2900	2007	69.207
2398	1	3	1	0	1000	2900	2007	69.207
2399	1	2	1	0	900	3200	2264	70.750
2400	1	3	1	0	900	3200	2264	70.750

profiles should be applied. This would improve the memory efficiency of the search algorithm [15].

5   Conclusion

In this paper, a two-phase forensic search procedure is described that applies q-gram distance in the pre-selection phase, in which the fragments of the examined data that deserve a more focused search are selected. The goal of the use of this kind of distance is to replace the constrained edit distance that is usually applied in such procedures but is computationally complex to compute. The q-gram distance is known to approximate well the *unconstrained* edit distance and its computational complexity is linear in the sum of the lengths of the involved strings. Behaviour of the q-gram distance when it replaces the constrained edit distance with special constraints regarding the maximum lengths of runs of deletions and insertions was studied experimentally over a 30 kB text file. Experimental results show that it is possible to achieve very high accuracy with the use of the q-gram distance for relatively large number of combinations of distance computation parameters. This indicates that q-gram distance might replace the constrained edit distance for general forensic search.

References

1. Beebe, N., Clark, J.: Dealing with terabyte data sets in digital investigations. In: Advances in Digital Forensics: Proceedings of the IFIP International Conference on Digital Forensics, pp. 3–16 (2005)
2. Boyer, R., Moore, J.: A fast string searching algorithm. Comm. ACM 20(10), 762–772 (1977)
3. http://www.dtsearch.com
4. Kashyap, R., Oommen, B.: The Noisy Substring Matching Problem. IEEE Trans. Software Eng. SE-9(3), 365–370 (1983)
5. Knuth, D., Morris, J., Pratt, V.: Fast pattern matching in strings. SIAM J. Computing 6(2), 323–350 (1977)
6. Kurtz, S.: Approximate String Searching under Weighted Edit Distance. In: Proceedings of Third South American Workshop on String Processing, Recife, Brazil, August, pp. 156–170 (1996)
7. Mihov, S., Schulz, K.U.: Fast approximate search in large dictionaries. Computational Linguistics 30(4), 451–477 (2004)
8. Mihov, S., Mitankin, P., Schulz, K.U.: Fast selection of small and precise candidate sets from dictionaries for text correction tasks. In: Proceedings of ICDAR 2007, vol. 1, pp. 471–475 (2007)
9. Oommen, B.: Recognition of Noisy Subsequences Using Constrained Edit Distances. IEEE Trans. Pattern Anal. Mach. Intell. PAMI-9(5), 676–685 (1987)
10. Roussev, V., Richard III, G.: Breaking the performance wall: the cases for distributed digital forensics. In: Proceedings of the Digital Forensics Research Workshop, pp. 1–16 (2004)
11. Petrović, S., Franke, K.: Improving the Efficiency of Digital Forensic Search by Means of the Constrained Edit Distance. In: Proceedings of the Third International Symposium on Information Assurance and Security, pp. 405–410 (2007)
12. Sellers, P.: The theory and computation of evolutionary distances: pattern recognition. Journal of Algorithms 1(4), 359–373 (1980)

13. Shi, F.: Fast Approximate Search in Text Databases. In: Li, Q., Wang, G., Feng, L. (eds.) WAIM 2004. LNCS, vol. 3129, pp. 259–267. Springer, Heidelberg (2004)
14. Shi, F., Mefford, C.: A New Indexing Method for Approximate Search in String Databases. In: Proceedings of Fifth International Conference on Computer and Information Technology (CIT 2005), pp. 70–76. IEEE Computer Society Press, Los Alamitos (2005)
15. Ukkonen, E.: Approximate string-matching with q-grams and maximal matches. Theoretical Computer Science 92, 191–211 (1992)

Similarity Visualization for the Grouping of Forensic Speech Recordings

Klara A. Weiand[1], Jos S. Bouten[2], and Cor J. Veenman[1,2]

[1] Intelligent Systems Lab,
University of Amsterdam, Amsterdam, The Netherlands
[2] Digital Technology & Biometrics Department,
Netherlands Forensic Institute, The Hague, The Netherlands

Abstract. In a forensic phone wiretapping investigation, a major problem is to get the full picture of the speakers involved. Typically, the wiretapped speech recordings are grouped using a clustering tool. The main disadvantage of such an approach is that in a bootstrapped scenario grouping errors accumulate. In this paper, we propose a visual approach to find similar speech recordings that probably stem from the same speaker. We first model the speech recordings and define suitable similarity measures between recordings. Then, through an approximate 2-D visualization of the inter-speech, similarities the investigator can identify clear groups of recordings and recordings that are harder to differentiate. We did extensive experiments on phone data of 50 speakers with 2 recordings per speaker. We tested quality of the 2-D visualization in relation to original high dimensional similarities. It turned out that for the original high dimensional similarity measure the nearest recording is almost always the one from the same speaker. In the 2-D visualization, we achieved that on average for all speech recordings a recording of the same speaker is among the 10 nearest recordings.

1 Introduction

Forensic speaker recognition is required is several situations, such as during an investigation to get the social network around a target, or when a questioned recording is compared with a suspect in a case. Since voice is in part a behavioral characteristic, it is not very constant and easy to alter. Also, emotional distress, health and simply aging can alter the characteristics of a speaker's voice significantly. Further, low-quality recordings which introduce noise specific to the recording and transmission device add to this problem.

While automatic speaker recognition systems have been reported to outperform naive human speaker recognition when the telephone handset is identical for all recordings and the data comes from the same recording session, automatic systems generally are far more sensitive to variation and noise than humans when conditions are mismatched [1] [2] [3]. This is attributed to the fact that humans use high-level cues while automatic speaker recognition systems operate at a lower, spectral level, where both kinds of variability have a bigger impact [3].

Though in forensics recordings made under good, constant conditions are rarely found, automatic forensic speaker recognition is certainly useful. First, neither of the cited studies mentions using normalization techniques to remedy the session and

S.N. Srihari and K. Franke (Eds.): IWCF 2008, LNCS 5158, pp. 169–180, 2008.

channel effects, let alone state of the art techniques. While intra-speaker variability still constitutes a problem and is an important focus of research, automatic speaker recognition performance under adverse conditions has steadily improved in recent years [4] [5] and can be expected to continue to do so. Secondly, automatic speaker recognition has many advantages over speaker recognition performed by a human expert. Due to human's sensitivity to language and reliance on high-level features, it is of importance for speaker recognition to be performed by native speakers. If the language is uncommon and its speakers in an area or country thus are more likely to be acquainted or related, it may not be possible to find an unbiased, willing person with enough expertise in speaker recognition to reliably perform the task. Automatic speaker recognition systems are mostly insensitive to problems stemming from language choice. Additionally, automatic speaker recognition systems can process the large amounts of data that often come with forensic casework and where time might be tight.

Whether the speaker recognition is done by an automatic system or by human expert, recognition errors have to be accounted for. That is, in practical forensic situations perfect recognition is an illusion. For evidence evaluation, i.e. when a suspects voice is compared to a questioned recording, the use of a matching score – the likelihood ratio (e.g. [6]) – is therefore becoming more and more accepted. Also in investigation cases, however, the reality of imperfect recognition has to be recognized. Traditional speaker recognition systems work independently of human interaction, taking a number of speech files as an input and giving the judgment on the identities – often augmented with confidence scores – as an output. The current systems do not aim at providing insight into the decision process and do not make the overall similarity relations accessible to human interpretation in an intuitive way. However, as described, human and automatic speaker recognition are complementary in their characteristics. It is therefore desirable to combine the human and automatic speaker recognition process and their respective strengths into a system where human-computer cooperation can lead to mutual improvement. Some approaches to integrating human interaction in automatic speaker recognition exist, but the human contribution here consists merely in feature preselection and the systems require the user to be expert phoneticians [7].

In this paper, we deal with the explicit modeling of uncertainties is the recognition of matching recordings. We propose a visual approach to find similar speech recordings that probably stem from the same speaker. We first model the speech recordings and define a suitable similarity measure between recordings. Through an approximate 2-D visualization of the inter-speech similarities, the investigator can identify clear groups of recordings and recordings that are harder to assign. In the next section, we first describe the problem statement we deal with in this research. In the following section, we describe the components that our system is built up from. In the Section Experiments, we report the elaborate experiments we did to test the plain speaker recognition performance and the effectiveness of the visualization.

2   Problem Statement

The problem we deal with is the presentation of similarities between speech recordings on a 2-D display. The application scenario is a wiretapping investigation over the phone with up to 50 speakers. The goal is to present the similarities between speech recordings such that: if there is more than one recording of a speaker, then for all these recordings the nearest recording on the 2-D display is from the same speaker. If there is only one recording of a speaker, then no constraints apply.

3   System Components

In this section, we describe the different components of our system. The system starts from a number of audio files, each of which contains speech from only one speaker, and displays the similarity relationships between the speakers in the individual audio files in two dimensions.

At first, the speech signal is segmented into overlapping slices known as frames. We use a common frame length of 30ms with 25% overlap between frames. Then, the frames are classified as being voiced or unvoiced. For the voiced sections we compute appropriate features and learn speech models. In order to compare speech models we need a mechanism to score the similarity between learned models. Finally, all similarity scores are put in a similarity matrix, which will be transformed to several 2-D visualizations. In the next subsection we describe these steps into some detail.

3.1   Extraction of Voiced Sections

Several studies that researched which parts of the speech signal contribute to good performance in speaker recognition systems found that using only the voiced sections yielded better results than using the complete signal or only removing the silent parts [8], [9] and thus the extraction of voiced sections is common in many speaker recognition systems. In our system, we use the combination of short time energy, autocorrelation, and zero crossing rate to establish that a frame is voiced. If all of these measures indicates that a frame is voiced it is flagged as voiced.

3.2   Speech Modeling

For the representation of the speech signals in the segmented frames, we use the Mel-Frequency Cepstral Coefficients (MFCC) [10] [11], that are widely used in current speaker and speech recognition systems [12]. To compensate for channel and session effects, Cepstral Mean Subtraction, (CMS) [13] [14] [15], is applied on the feature matrix by subtracting the mean of each feature dimension from all feature values in that dimension.

For the modeling of the speech recordings, we abstract the set of 18-D MFCC vectors by estimating their 18-D distribution. Commonly, Gaussian Mixture Modeling (GMM) is used for this purpose. In this study we compare two different methods to estimate the Gaussian mixtures. First, we use Expectation Maximization (EM) [16] to optimise the weights, means and covariances of the Gaussians in the mixture. Second, we estimate the mixture by Maximum A Posteriori (MAP) adaptation [17] starting from an EM-trained background GMM, also called an Universal Background Model (UBM) [18].

Since in our case, the MAP adapted speaker models only differ in their means but have identical covariance matrices and weights, the mean vectors are sufficient to fully describe the speaker and feature-based classification can be applied. A concatenation of all mean vectors in one GMM yields a so-called supervector [19]. For example, if a 128-mixture GMM is adapted using 18-dimensional MFCC data, a supervector with 2304 dimensions results. In principle, any feature based classifier can be applied to learn a speaker classifier. The Support Vector Machine (SVM) has been applied successfully in this domain [19], where the following linear kernel is commonly used:

$$K(x_1, x_2) = \Sigma_{i=1}^{N} (\sqrt{w_i} \Sigma^{-\frac{1}{2}} x_i^1)' \cdot (\sqrt{w_i} \Sigma^{-\frac{1}{2}} x_i^2) \tag{1}$$

where w and Σ are the respective weights and covariances of the GMM mixture component. In this case, the diagonal scaling of the means can also be computed on the supervectors beforehand, which means that a regular linear kernel can be used.

An SVM classification model for a certain target speaker is then learned by representing the target class (speaker) through one supervector and a cohort speaker supervectors, which resembles the background model.

3.3 Speech Similarity Measure

Once we have the models of the individual speakers, we need a measure to determine how similar or dissimilar the models and their original data vectors (MFCC) are from each other.

Similarities for SVMs. The distance between a trained SVM classifier and a supervector is determined by simply using the classifier on the supervector. The resulting value, the distance from the boundary, indicates the relative fit for both classes on a continuous scale, thus yielding a number that quantifies how close to the speaker on which the SVM was trained the new data lies relative to the distance of the cohort.

Similarities for GMMs. The GMM distributions could be compared directly, but measures for the distance between distributions like the Bhattacharyya distance or the chi square metric often impose constraints on the characteristics of the distributions or are expensive to compute. Alternatively, the likelihood expresses the relation between the GMM created from one recording and the data vectors created from another speech recording. Since the GMM components are multiple ordinary normal distributions, the likelihood can be computed for each mixture component on the basis of the general Gaussian likelihood function using the respective means and covariance matrices. These values are summed and, to represent the contribution of each Gaussian to the GMM correctly, weighted, leading to the following definition for the likelihood function for GMMs [16]:

$$p(\overrightarrow{x}|\lambda) = \Sigma_{i=1}^{M} w_i p(\overrightarrow{x}|\lambda_i). \tag{2}$$

The likelihood of a GMM is thus simply a weighted linear combination of the densities of its component distributions.

Note that above equation will yield one value per MFCC vector. Each MFCC vector represents only a fraction of a second of speech and consequently there will be a big number of likelihood values that has to be combined into a single score. Since the lengths of the MFCC vectors vary depending on the length of the voiced sections in the original utterance, the mean of the log likelihood values is used:

$$log(p(X|\lambda)) = \frac{1}{T}\Sigma_t log(p(\overrightarrow{x}|\lambda)) \tag{3}$$

An alternative used in speaker recognition is the Bayesian Information Criterion (BIC) [20]. It is based on the log likelihood score, but additionally penalises the model complexity. Since in our case the complexity of the model is fixed and the number of MFCC vectors per recording is similar, the added value of the BIC score is limited. We do not consider it further in this study.

The likelihood score as described lacks one important characteristic compared to the SVM scores: They are not normalized with respect to the cohort or background model that represents the average speaker. This means that the score only represents the relative distance between a model and some data, but this score does not relate to how likely an average, nondescript speaker is to produce an utterance that leads to the same feature vectors and therefore how significant the original score is.

This problem is easily remedied by redefining the score as a ratio of likelihoods scores [17]:

$$ll_{ratio} = \frac{log(p(X|\lambda))}{log(p(X|\lambda_{background}))} \tag{4}$$

The likelihood ratio thus expresses the magnitude of the score relative to the score of the background speaker, making scores comparable across the evaluation of the same data for different models.

Symmetric Similarities. Ideally, both described similarity measures are symmetric. That is, given two sets of data (MFCC) vectors X_1 and X_2, the similarity $s(X_1, X_2) = s(X_2, X_1)$, where $s(X_i, X_j)$ is computed through SVM or GMM modeling. Because the data per speaker is limited and the modeling is imperfect, in practice the similarity is not symmetric. Since this is undesirable for the visualization, we make the similarity symmetric as follows:

$$s'(X_1, X_2) = s(X_1, X_2) + s(X_2, X_1) \tag{5}$$

3.4 Visualization

For the visualization of the similarities between speaker models or between models and data vectors from recordings, we need to transform the original similarities to similarities or distances in 2-D. In the following, two dimension reduction algorithms used in the experiments, Multidimensional Scaling [21] and Isomap [22] will be described. These non-linear dimension reduction techniques were chosen, since on the one hand it is an established technique (MDS) and on the other hand a more recent related technique (Isomap). The similarities between the two dimension reduction algorithms will be explained in the following.

Multidimensional Scaling. Multidimensional Scaling is an established set of techniques that find wide application in various scientific fields like Psychology and Economy. It takes a distance matrix and tries to find a configuration of points in lower dimensional space whose Euclidean distances match the distances in the input matrix.

There are multiple variants of MDS, here a nonlinear variant using a Sammon mapping which is performed on a symmetric distance matrix with zero diagonal is described and evaluated.

The MDS procedure consists in the minimization of an error function, the stress. We consider the original metric MDS with the following optimization or stress function:

$$Stress = \Sigma_{i=I}^{N}[d(x_1, x_2) - d'(x_1, x_2)]^2 \tag{6}$$

where $d(x_1, x_2)$ is the distance between x_1 and x_2 in the original space and $d'(x_1, x_2)$ is their distance in the mapped space.

Isomap. The Isomap algorithm [22] extends MDS by first calculating the geodesic distances in the high-dimensional space and then applying metric MDS on the resulting distance matrix. The idea behind the nonlinear dimension reduction technique Isomap is to find coordinates in lower dimensional space that preserve the intrinsic geometry of the data by finding the underlying manifold, a topological space that is locally but not globally Euclidean. Due to this property of the manifold, the distance between adjacent points is calculated using the Euclidean distance, while the distance between points that are far away from each other, the geodesic distance is calculated. This is done by first constructing a graph in which nearby points are connected via edges, and then calculating the distance between the two points along the shortest path between them.

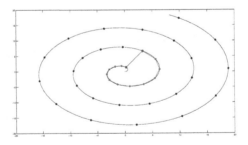

Fig. 1. Geodesic and Euclidean distance illustrated

As an illustration, consider the dataset in figure 1 where the circles indicate the data points. The direct distance between the two red points, indicated by the red line, ignores the underlying spiral form of the data and the Euclidean distance calculated between the points. The geodesic distance on the other hand can be approximated by building a graph structure where each point is connected to its nearest neighbor and then summing the distances between the edges of the shortest path, here indicated by the magenta line.

The geodesic distance here is the sum of the Euclidean distances between the magenta circles and the first and last red circle.

4   Experiments

To test the speech modeling and the visualization that is based on it, we did extensive experiments. Below, we first describe the dataset used and in the following section the results.

4.1   Dataset

In absence of true forensic wiretapped recordings, all tests were run on data from the SwitchBoard-2 phase 1 corpus[1] which is frequently used in the NIST speaker recognition task. Each file had speech from only one speaker, meaning that speaker segmentation was not needed, and contained a concatenation of consecutive turns recorded during a telephone conversation.

The phone conversations were collected by the Linguistic Data Consortium, the speakers are all US Americans, mostly from the Mid-Atlantic area of the US and college students. Conversations lasted about 5 minutes in total, a proposed topic for each conversation was given but not enforced. As is typical for telephone data, the audio was encoded at a sampling rate of 8KHz and 8 bits per sample. All files were initially about 60 seconds in duration, of which roughly a third was discarded when removing the unvoiced sections, silence and noise.

There were three segments per speaker, two of them from the same session. The third segment was recorded during a separate sessions but the same handsets were used. We do not report the mixed session results, because of space limitations. Also, for the visualizations there were no systematic differences between the methods with the same session results that we report here.

172 files, 86 female and 86 male recordings, were separated from the rest of the files and used as data for the computation of the cohorts and background models which were thus balanced in terms of gender [17].

Performance measure. The wiretapping scenario, for which our system is designed, typically has up to 50 speakers per case. Therefore, we randomly drew the same session training and test recordings of groups of 50 speakers from the 162. We repeated this 50 times and averaged the results to become less sensitive to a specific sample of 50 speakers. Moreover, random effects in the GMM estimation and MDS optimization are averaged out too.

To quantify how well the recordings of the same speaker are found as nearest recording by the system (as required in the Problem Statement), we propose the nearest neighbor score. This is a score that counts for how many recordings the recording of the same speaker is among the k-nearest recordings. In other words, the nearest neighbor score (NN-score) is a function of k. In our setting, there are in total models of 100 recordings. Then, at $k = 100$ the score becomes always 1. Further, we define the Nearest Neighbor Hit-score (NNH-score) as the k value for which the NN-score reaches 1.

---

[1] http://www.ldc.upenn.edu/Catalog/CatalogEntry.jsp?catalogId=LDC98S75

Table 1. Table with Nearest Neighbor Hit-scores in original high-dimensional space and after dimension reduction

Method	High dimensional	MDS	Isomap			
			3	7	10	15
EM-GMM	1.00	23.50	12.36	18.06	11.34	18.28
MAP-GMM	1.00	46.26	8.80	12.46	34.74	13.52
SVM	1.00	23.14	6.60	49.32	17.74	18.32

4.2 Results

For each type of model, one distance matrix for the best performing number of mixtures and for each session condition was selected and the dimension reduction algorithms were applied. For the EM-GMM, 32 mixture components were chosen, while for the SVM and MAP-GMM, distance matrices created from 256 mixture component models were selected.

Besides the target number of dimensions, the MDS algorithm has no parameters. Isomap on the other hand additionally has as parameter the number of neighbors for the underlying nearest neighbor graph. In the experiments we set this parameter to 3,

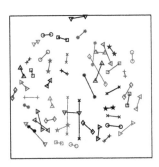 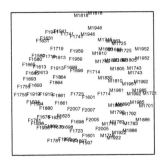

(a) MDS

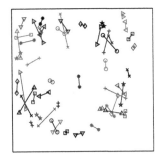 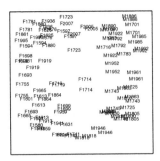

(b) Isomap (10 neighbors)

Fig. 2. Visualization of the speaker similarities using EM-GMM modeling

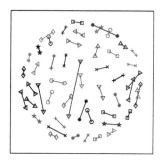 

(a) MDS

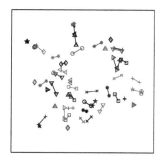 

(b) Isomap (3 neighbors)

Fig. 3. Visualization of the speaker similarities using MAP-GMM modeling

7, 10 and 15 neighbors, respectively. Moreover, as the MDS and Isomap models are optimized nondeterministically, every experiment was run 50 times and the result of the run with the lowest overall remaining stress was chosen.

Table 1 gives the mean NNH-scores (averaged over 50 samples) for the models in the original high-dimensional space and after dimension reduction through MDS and Isomap. The table clearly shows that for same session conditions, the system has close to perfect average performance. Also, MDS is outperformed by Isomap for all modelings. The neighborhood size for Isomap has an effect on performance, but there is no clear relation. The type of model appears to have no systematic influence on the performance of the different dimension reduction techniques. Isomap(3) has the best performance for MAP-GMM and SVM modeling. Also for EM-GMM this dimension reduction scheme is close to the best.

Visualizations. For each model type, 2-D plots representing the speaker distances after dimension reduction with each of the techniques were created. In the case of Isomap, the neighborhood size yielding the lowest NNH-score was used.

Two types of plots were created, one displaying each speech file as a symbol connected to the speaker's other speech file in the set by a line. The other plot type shows

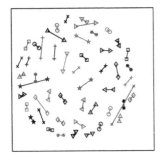 

(a) MDS

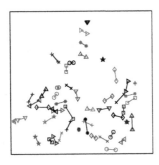 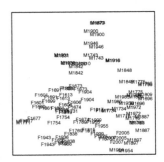

(b) Isomap (3 neighbors)

Fig. 4. Visualization of the speaker similarities using SVM modeling

the name of the speech files with the prefix, 'M' or 'F', indicating a male and female speaker respectively, and where identical numbers mean that two files came from the same speaker. The plots in figure 2(a) show the MDS visualization where the speaker models are 32-mixture EM-estimated GMMs.

Remarkably, gender classification emerges although all steps of processing are gender-neutral or balanced in gender. That is, the recordings of male and female speakers are clearly separated in the 2-D plot, while the gender of the speakers is not disclosed at any stage in the system.

Figure 2(b) shows the Isomap visualizations. These visualizations show many cases where matching models are very close together, but also a number of cases where files from the same speaker have a relatively large distance between them. All in all, the distances in this plot are more varied than those in the MDS plot.

Figure 3(a) and 3(b) show the speaker spaces resulting from the application of the two dimension reduction techniques to distance matrices created from 256-mixture MAP-adapted GMMs. In the MDS plot, again, apart from a small number of outliers, distances between matching models are small and relatively similar. As the label plot shows, speakers here are not automatically arranged according to gender, but rather in several groups with the same gender.

As in the other two models, the MDS visualization (figure 4(a)) based on distances determined by SVM classification shows relatively small, constant distances between matching files with few big outliers in terms of distance. The models again are separated by gender.

5 Conclusion

In this paper, we proposed a system for visualizing the similarities between forensic speech recordings. The system explicitly models and shows the uncertainty in the speaker recognition process, so that the user can decide on the basis of the visualisation which recordings stem from the same speaker. To this end, the user may apply additional knowledge.

In order to be able to visualize speech similarities we built several speech models that are based on Gaussian Mixtures and defined similarity measures between models. Finally, we applied two dimension reduction techniques for visualization; Multi Dimensional Scaling (MDS) and Isomap, which is based on a nearest neighbor graph.

We conducted extensive experiments using the SwitchBoard-2 phase 1 corpus, which showed that the speaker recognition performance in same session conditions was close to perfect. Though this is not realistic, such a base-line shows what performance we can obtain under good conditions. This then also holds for the dimension reduction and visualization. Although none of the dimension reduction techniques was able to represent the distances found in high-dimensional space, it was found that the right choice of dimension reduction technique leads to two-dimensional visualizations that display the speaker similarities with high accuracy. The MDS technique in general shows a few outliers for which the matching speech recording is far away. With the Isomap technique this hardly happened, so from this study we conclude that Isomap is the most suitable. This also follows from the Nearest Neighbor Hit-scores that we introduced in this paper.

Interestingly, representations were arranged by gender in most visualizations, while the gender of the speakers was not disclosed to the system. This further indicates that the visualization display meaningful speaker similarities.

Finally, further tests are needed with more realistic data and involving forensic investigators in the interpretation of the visualizations.

References

1. Schmidt-Nielsen, A., Crystal, T.H.: Human vs. machine speaker identification with telephone speech. In: Proceedings of the 5th International Conference on Spoken Language Processsing, Sydney, Australia (1998)
2. Ezzaidi, H., Rouat, J.: Speaker identification by computer and human evaluated on the SPIDRE corpus. Canadian Acoustics 28(3) (2000)
3. Alexander, A., Dessimoz, D., Botti, F., Drygajlo, A.: Aural and automatic forensic speaker recognition in mismatched conditions. Forensic Linguistics: The International Journal of Speech, Language and the Law 12(2) (2005)

4. Przybocki, M.A., Martin, A.F., Le, A.N.: NIST speaker recognition evaluation chronicles – part 2. In: Proceedings of the Odyssey 2006. Speaker and Language Recognition Workshop, San Juan, Puerto Rico, pp. 1–6 (2006)
5. Martin, A.F., Przybocki, M.A.: The NIST speaker recognition evaluations: 1996-2001. In: Proceedings of the Odyssey 2001. Speaker and Language Recognition Workshop, Crete, Greece (2001)
6. Gonzalez-Rodriguez, J., Drygajlo, A., Ramos-Castro, D., Garcia-Gomar, M., Ortega-Garcia, J.: Robust estimation, interpretation and assessment of likelihood ratios in forensic speaker recognition. Computer Speech and Language 20, 331–355 (2006)
7. Bimbot, F.B., Fredouille, J.F., Gravier, C., Magrin-Chagnolleau, G., Meignier, I., Merlin, S., Ortega-Garcia, T., Petrovska-Delacretaz, J., Reynolds, D.A.: A tutorial on text-independent speaker verification. EURASIP Journal on Applied Signal Processing, 430–451 (2004)
8. Kim, J.K., Shin, D.S., Bae, M.J.: A study on the improvement of speaker recognition system by voiced detection. In: Proceedings of the 2002 45th Midwest Symposium on Circuits and Systems, Tulsa, USA, vol. 3, pp. 324–327 (2002)
9. Krishnamachari, K., Yantorno, R.: Spectral autocorrelation ratio as a usability measure of speech segments under co-channel conditions. In: Proceedings of the IEEE International Symposium on Intelligent Signal Processing and Communication Systems, Honolulu, USA (2000)
10. Davis, S., Mermelstein, P.: Comparison of parametric representations for monosyllabic word recognition in continuously spoken sentences. IEEE Transactions on Acoustics Speech and Signal Processing 28(4) (1980)
11. Campbell, J.: 8: Speaker Recognition. In: Biometrics - Personal Identification in Networked Society. Springer, Heidelberg (2002)
12. Reynolds, D.A.: An overview of automatic speaker recognition technology. In: Proceedings of the IEEE International Conference on Acoustics, Speech, and Signal Processing, Orlando, USA, pp. 4072–4075 (2002)
13. Atal, B.: Effectiveness of linear prediction characteristics of the speech wave for automatic speaker identification and verification. Journal of the Acoustical Society of America 55(6), 1304–1312 (1974)
14. Furui, S.: Speaker Recognition. In: Survey of the State of the Art in Human Language Technology, pp. 36–41. Cambridge University Press, Cambridge (1998)
15. Johnsen, M.H., Svendsen, T., Harborg, E.: Experiments on cepstral mean subtraction (CMS) and rasta-filtering applied to SAMPA phoneme recognition. In: COST 249, Nancy, France (1995)
16. Reynolds, D.A., Rose, R.: Robust text independent speaker identification using Gaussian mixture speaker models. IEEE Transactions on Speech and Audio Processing 3(1), 72–83 (1995)
17. Reynolds, D.A., Quatieri, T.F., Dunn, R.B.: Speaker verification using adapted Gaussian Mixture Models. Digital Signal Processing 10, 19–41 (2000)
18. Barras, C., Gauvain, J.L.: Feature and score normalization for speaker verification of cellular data. In: Proceedings of the IEEE International Conference on Acoustics, Speech, and Signal Processing, Hong Kong, China, vol. 2, pp. 49–52 (2003)
19. Campbell, W.M., Sturim, D.E., Reynolds, D.A.: Support Vector Machines using GMM supervectors for speaker verification. IEEE Signal Processing Letters 13(5), 308–311 (2006)
20. Schwartz, G.: Estimation of the dimension of a model. Annals of Statistics 6, 461–464 (1978)
21. Torgerson, W.S.: Theory and methods of scaling. John Wiley, New York (1958)
22. Tenenbaum, J.B., de Silva, V., Langford, J.C.: A global geometric framework for nonlinear dimensionality reduction. Science 290(5500), 2319–2323 (2000)

Handwritten Signature and Speech: Preliminary Experiments on Multiple Source and Classifiers for Personal Identity Verification

Donato Impedovo[1,3], Giuseppe Pirlo[2,3], and Mario Refice[1]

[1] Dipartimento di Elettrotecnica ed Elettronica, Politecnico di Bari,
via Orabona 4, 70125 Bari (I)
impedovo@deemail.poliba.it, refice@poliba.it
[2] Dipartimento di Informatica, Università degli Studi di Bari,
via Orabona 4, 70126 Bari (I)
pirlo@di.uniba.it
[3] Centro "Rete Puglia", Università degli Studi di Bari,
via G. Petroni 15/F.1, 70124 Bari (I)

Abstract. This paper presents a personal verification system based on two different biometric traits: handwritten signature and speech. The signature verification system uses contour-based features and a Dynamic Time Warping technique for matching. The speaker verification system uses cepstral based coefficients and is based on a Hidden Markov Model statistical classifier. In the decision combination stage, the decisions provided by the two systems are combined according to a simple abstract–level combination approach. The experimental results related to a real-scenario demonstrate the effectiveness of the proposed approach and highlight some profitable directions for further developments.

Keywords: Biometry, Personal Authentication, Signature Verification, Speaker Verification, Multi-expert system.

1 Introduction

In modern society there is an increasing request for more and more reliable personal verification systems. In fact, personal verification plays a very important role not only in personal security but also in data protection and transaction validation [1].

In this context, biometry offers potential for recognizing an individual on the basis of his/her individual characteristics and it is expected to provide better security solutions than traditional identification methods, based on physical devices (ID cards, badges, etc.) or information (passwords, PINs, etc.). In a biometric system, both physiological or behaviour traits can be considered, depending on the specific requirements of the applications, acceptance by users etc. [2,3].

S.N. Srihari and K. Franke (Eds.): IWCF 2008, LNCS 5158, pp. 181–191, 2008.

This paper presents a personal verification system based on multi-modal biometrics. Personal verification is performed by combining the results of a hand-written signature verification system and a speaker verification system.

Hand-written signature verification is one of the most interesting biometric techniques, since signature is one of the most widespread means for personal verification and it is well-accepted by users [4,5,6,7]. Therefore, signature verification is a field of intensive research that concerns not only relevant scientific challenges, but also valuable commercial applications, like those related to banking transactions, health care and e-commerce [1,8,9,]. Although research on hand-written signature verification has a long history, recently, along with the growth of internet, the field of automatic signature verification has been considered with a renewed interest [4,10,11]. Hand-written signature verification concerns two different approaches: static (or off-line) or dynamic (on-line) verification. In the first case, signature acquisition occurs after specimens have been written. In this case no dynamic information is available for verification aims. Conversely, when dynamic signatures are considered, signatures are acquired during the writing process. Therefore, a wide set of dynamic information can be used for verification aims [5,6]. In the system here presented, signature verification is performed by Dynamic Time Warping (DTW), that is used to match simple contour-based functions.

On the other hand Speaker Verification (SV) plays a very important role in the biometric issue. Speaker verification is the task of deciding if the speaker is who he claims to be [12,13]. Speaker verification includes at least two different branches: text independent and text dependent applications. The first intends to verify the identity using any speech emission from the claimant, while the second approach is based on the use of a specific utterance. Since the signature verification deals with the recognition of a pattern that has a meaning content not varying over time, for the speaker verification system, a text-dependent application has been considered here for this preliminary investigation. These kinds of systems are, nowadays, the ones with the highest performance and can be applied successfully in real situations [14]. In the system here presented, speaker verification is performed by the use of a statistical classifier – Hidden Markov Model (HMM) – and cepstrum coefficients as features.

The paper is organized as follows: Section 2 deals with system description. Sections 3 and 4 describe the Signature Verification System and the Speaker Verification System, respectively. Section 5 describes the combination technique. Some experimental results are summarized in Section 6. Section 7 presents the conclusion of the paper and briefly describes some profitable areas for further developments.

2   System Description

Figure 1 shows schematically the personal verification system presented in this paper. It consists of three main modules: a) Signature Verification Module; b) Speaker Verification Module; c) Decision Combination Module.

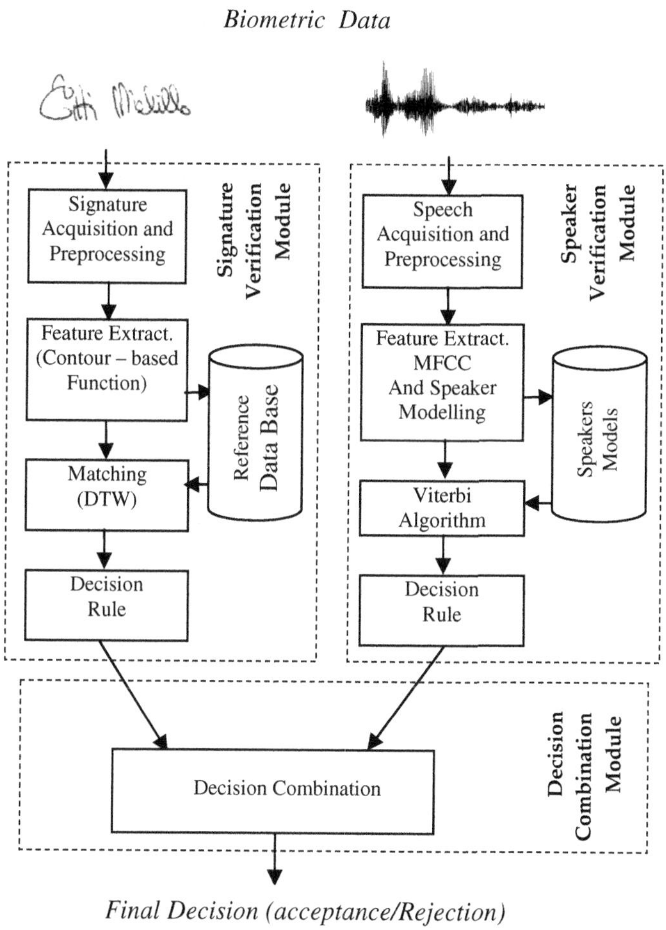

Biometric Data

Fig. 1. The Personal Verification System

3 Signature Verification Module

In this paper a static signature verification system has been considered. It uses the upper and lower contour as function features and adopts Dynamic Time Warping (DTW) for signature matching. Each signature S is scanned and normalised so that the rectangle surrounding it is scaled to a fixed area [15]. Successively, the upper profile (PU) and lower profile (PL) are extracted [16, 17]. Figure 2 shows an input signature (Fig. 2a) and the upper (Fig. 2b) and lower (Fig. 2c) contours.

The comparison process is performed by a DTW algorithm, based on Euclidean Distance. Specifically, if the upper profile is considered, let

$$P^U_1 : (z^U_1(1), z^U_1(2), ..., z^U_1(i_k), ..., z^U_1(M_1)) \qquad (1)$$

be the upper profile of S_1 (of length equal to M_1) and

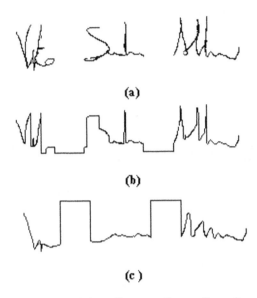

Fig. 2. Hand Written Signature: Feature Extraction

$$P^U_2 : (z^U_2(1), z^U_2(2), \ldots, z^U_1(j_k), \ldots, z^U_2(M_2)) \tag{2}$$

be the upper profiles of S_2 (of length equal to M_2). A warping function between PU_1 and PU_2 is defined as any sequence of couples of indexes identifying points of PU_1 and PU_2 to be joined [18]:

$$W(P^U_1, P^U_2) = c_1, c_2, \ldots, c_K, \tag{3}$$

where $c_k = (i_k, j_k)$ $(k, i_k, j_k$ integers, $1 < k < K$, $1 < i_k < M_1$, $1 < j_k < M_2)$.
Now, if the Euclidean Distance is considered $d(c_k) = d(z^U_1(i_k), z^U_2(i_k))$ between points of P^U_1 and P^U_2, associated to the warping function $W(P^U_1, P^U_2)$, it is possible to compute the dissimilarity measure

$$D_{W(P^U_1, P^U_2)} = \sum_{k=1}^{K} d(c_k) \tag{4}$$

The elastic matching procedure detects the warping function $W^*(P^U_1, P^U_2) = c^*_1, c^*_2, \ldots, c^*_K$. which satisfies the monotonicity, continuity and boundary conditions, and for which it results [18, 19]:

$$D_{W^*(P^U_1, P^U_2)} = \min_{W(P^U_1, P^U_2)} D_{W(P^U_1, P^U_2)}. \tag{5}$$

During the training process, the features extracted from the set of reference signatures are enrolled into the personal database. In the comparison phase, they are matched against those belonging to the input (test) signature. The result is used to judge the authenticity of the input signature.

Precisely, in our system three genuine specimens have been considered for reference [20,21]:

- $S_{r,1}$, described by $P^U_{r,1}$ and $P^L_{r,1}$;
- $S_{r,2}$, described by $P^U_{r,2}$ and $P^L_{r,2}$;
- $S_{r,3}$, described by $P^U_{r,3}$ and $P^L_{r,3}$.

From the analysis of the three specimens, the personal variability of the signer is estimated as:

- $T^U = \max D(P^U_{r,i_1}, P^U_{r,i_2})$, with $i_1, i_2 \in \{1,2,3\}$ $(i_1 \neq i_2)$, that estimates the maximum personal variability for the upper profile;
- $T^L = \max D(P^L_{r,i_1}, P^L_{r,i_2})$, with $i_1, i_2 \in \{1,2,3\}$ $(i_1 \neq i_2)$, that estimates the maximum personal variability for the upper profile.

Now, let S_t be a test signature, described by P^U_t and P^L_t. The verification of S_t is performed by matching it against $S_{r,1}$, $S_{r,2}$ and $S_{r,3}$. Specifically S_t is considered a genuine specimen if and only if both the following conditions are true:

1. $\min_{(for\ k=1,2,3)} D(P^U_t, P^U_{r,i_k}) \leq T^U$
2. $\min_{(for\ k=1,2,3)} D(P^L_t, P^L_{r,i_k}) \leq T^L$.

4 Speaker Verification Module

The spoken identity is acquired by a VAD (voice activation detection) to remove silence from recordings. After this step, the feature extraction is performed. Mel Frequency Cepstral Coefficients (MFCCs), their time derivatives and the respective energy parameter have been considered, they are obtained from the power spectrum of the signal. Since speech is a non stationary signal, in order to consider the Discrete Fourier Transform (DFT) a short time analysis is performed: the signal is framed into constant frame size (generally the duration is between 20 and 30ms, and a step of 10ms is adopted in the framing process). For each frame the DFT is computed:

$$X(k) = \sum_{n=0}^{N-1} w(n)x(n)\exp(-j2\pi kn/N) \tag{6}$$

for $k = 0,1,...,N-1$, where $x(n)$ is the time discrete signal in the frame with length N, k corresponds to the frequency $f(k) = kf_s/N$, f_s is the sampling frequency in Hertz and $w(n)$ is the Hamming time window given by $w(n)=0.54-0.46cos(\pi n/N)$.

The magnitude spectrum $|X(k)|$ is then scaled in frequency and magnitude. The frequency is scaled by Mel filter bank $H(k,m)$ and then the logarithm is considered:

$$X'(m) = \ln\left(\sum_{k=0}^{N-1} |X(k)| \cdot H(k,m)\right) \tag{7}$$

for $m = 1, 2,...., M$, with M the number of filter banks and $M << N$. The Mel filter bank is a collection of triangular filters defined by the center frequencies $f_c(m)$ and defined as follow:

$$H(k,m) = \begin{cases} 0 & for \quad f(k) < f_c(m-1) \\ \dfrac{f(k)-f_c(m-1)}{f_c(m)-f_c(m-1)} & for \quad f_c(m-1) \le f(k) \le f_c(m) \\ \dfrac{f(k)-f_c(m+1)}{f_c(m)-f_c(m+1)} & for \quad f_c(m) \le f(k) < f_c(m+1) \\ 0 & for \quad f(k) > f_c(m+1) \end{cases} \qquad (8)$$

and the center frequencies of the filter banks are spaced logarithmically on the frequency axis. Finally, the MFCCs are obtained by computing the Discrete Cosine Transform (DCT) of $X'(m)$:

$$c(i) = \sum_{m=1}^{M} X'(m)\cos\left(i\frac{\pi}{M}\left(m-\frac{1}{2}\right)\right) \qquad (9)$$

for $i = 1, 2, ..., M$, and $c(i)$ is the ith MFCC.

The Mel warping transforms the frequency scale to place less emphasis on high frequencies: it is based on the non linear human perception of the frequency sounds. For each frame, over the MFCC, the delta cepstrum coefficients (time derivates of the MFCC) and the respective power parameters have been also considered [22,23,24].

The verification phase is based on HMMs with continuous observation densities. A continuous HMM is able to keep information and to model not only the sound, but also the articulation and the temporal sequencing of the speech: the sequencing of sound (reported on the state's transitions probabilities) in the training data plays an important role, representing the sound sequences of the testing data [25,26]. The Hidden Markov Models considered in all the experiments in this paper adopt a left-to-right no skip topology. For each state, the Gaussian observation probability-density function (pdf) is used to statistically characterize the observed speech feature vectors.

An HMM λ can be characterized by a triple of state transition probabilities A, observation densities B, and initial state probabilities Π through the following notation:

$$\lambda = \{A, B, \Pi\} = \{a_{i,j}, b_i, \pi_i\} \qquad (10)$$

with $i, j = 1,...,N$, where N is the total number of states in the model and $a_{i,j}$ is the transition probability from the state i to j. Given an observation sequence (features vectors) $O=\{o_t\}$ with $t=1,...,T$, the continuous observation probability density for the state j is characterized as a mixture of Gaussian probabilities:

$$b_j(o_t) = \Pr(o_t \mid j) = \sum_{m=1}^{M} c_{jm} P(o_t; \mu_{jm}, R_{jm}) \qquad (11)$$

with

$$P(o_t; \mu_{jm}, R_{jm}) = (2\pi)^{-d/2} \mid R_{jm} \mid^{-1/2} \exp\left\{\frac{1}{2}(o_t - \mu_{jm})^T R_{jm}^{-1}(o_t - \mu_{jm})\right\} \qquad (12)$$

where M is the total number of the Gaussian components in the mixture, μ_{jm} and R_{jm} are the d-dimensional mean vector and covariance matrix of the mth component at state j and finally c_{jm} are the mixture weights which satisfy the constraint $\sum_{m=1}^{M} c_{jm} = 1$.

The mentioned model parameters have been estimated by the Baum-Welch iterative methods (expectation-maximization EM algorithm), in order to maximize $Pr(O|\lambda)$ [27, 28].

For each identity to be verified, two HMMs have been trained: the first represents the identity pronounced by the genuine speaker, while the second represents the identity claimed by impostors. Each model has 8 states and 3 Gaussian in the mixtures (this has been decided experimentally).

The recognition phase is based on the Viterbi algorithm [29]. Let λ_k be the model for the genuine k-th identity and λ^I_k the anti-model, given k the identity to be verified, the quantity S_k is computed as follow:

$$S_K = \log\left(\Pr(O \mid \lambda_k) - \Pr(O \mid \lambda^I_k)\right) \tag{13}$$

The anti-model λ^I_k, in literature, is also known as "world model" or "Universal Background Model" (UBM). Obviously in a real application there is no possibility to a-priori know the impostor's characteristic and to built a specific model for him/her: testing the system on impostor speakers belonging to the training set for the anti model would be misleading and would lead to biased results. For this reason impostor's trials in the testing phase are from outset training's ones.

The identity is considered from the genuine speaker or from the impostor according to a threshold value. The threshold could be speaker dependent or independent: in this work the second approach has been followed.

5 Decision Combination Module

Decision combination is a very important stage. In fact, the selection of the most profitable approach to be considered for decision combination is a critical task. Mainly, when verification responses are considered, decision combination can be performed at *abstract-level* , when the combination method uses the Boolean value provided by each verification module; or at *measurement-level* , when the combination method also uses the confidence value provided by each verification module [30, 31].

In general, *measurement-level* methods are expected to achieve better performance than *abstract-level* methods. Unfortunately, the combination of measures obtained by multiple verifiers requires the definition of complex normalization techniques [31]. Furthermore, in the field of multi-modal biometrics, the independence of the different traits ensures that a significant improvement in performance can be achieved by a multi-expert approach even when simple *abstract-level* combination rules are considered [33, 34].

According to these considerations, in this paper, three simple *abstract-level* rules have been adopted to combine the verification decisions provided by the Signature Verification Module and Speaker Verification Module [34]: the "AND rule", the "OR rule" and the "AND rule" with a re-trial option (i.e. an additional sample is required and verified in the case in which the two verifiers disagree).

6   Experimental Results

Twenty-two identities were considered for preliminary experiments. Speech recordings and signatures acquisition happened in four or five sessions, the number of signature/speech acquisition is not uniform among identities, this could be seen as an inconvenient, but it introduces realism into the database.

In the first session, each person provided three hand-written signatures and pronounced their respective "name surname" ten times. The three signatures were used as reference in the signature verification system. The ten speech recordings give a medium amount of 14.2 seconds of "speech" per person before the VAD, they were used to train the HMMs. In the first session speaker impostors were also enrolled for a total population of 54 persons. For each genuine identity an anti-model was trained on 3 different impostors pronouncing the real identity "name surname". Testing sessions were spaced over a time span of about three months: from 60 to 140 verification trials were performed for each person (the number of genuine trials is equal to the number of forgery trials). The signatures were written on a white sheet and successively acquired by a flat-bed standard scanner at 300dpi. Speech was acquired immediately after the signature by using a classic head-set microphone and sampled at 22KHz in an home/office environment. Successively, adopting a time windowing of 25ms updated every 10ms, 19 MFCC, their time derivatives and two energy values were extracted from speech files. The experimental setup tries to emulate real conditions were an exhaustive training cannot be performed and the number of access attempts from users is not uniform among them.

Performance were evaluated by considering the false rejection errors (type I errors, FR), caused by the rejection of genuine identities, and false acceptance errors (type II errors, FA), caused by the acceptance of impostors [1,2]. At the beginning of each session, impostors had the possibility to observe the genuine signature and to listen the spoken password, moreover they were present during the genuine user's trials. It has to be underlined once more that in no case speaker impostors used for the creation of a speaker's anti model was successively used to perform identification on that identity (too optimistic results on FA would be obtained). Table 1 reports the performance of the two stand alone systems.

Table 1. Performance for stand alone systems

	FR	FA
Speaker Identification System	5,27	2,08
Signature Verification System	20,08	54,71

Table 2 reports the overall performance of the combined system.

Table 2. Performance for stand alone systems

	AND	OR
Type I Error Rate (FR)	16,19	1,78
Type II Error Rate (FA)	1,47	42,61

The different trend in the FR and FA in the two cases is related to the fact that the speaker verification module, thanks to anti-models, makes the FA rate for the considerably lower than that observed for the signature verification one, in fact many forged signature appeared quite similar to the genuine ones, from this point of view dynamic features would improve the performance.

Since a relative high FR could be tolerated by genuine users for security reasons, the AND combination appears to be a good starting point. Successively the following decision policy has been adopted:

```
if(spv_out!=sgv_out)
{
          if (spv_out==0)
            spv_out_2=spvmod();
          else
            sgv_out_2=sgvmod();
}
```

where

- spv_out and sgv_out are the Boolean output respectively of the speaker and signature verification systems,
- 0 stands for rejection, 1 for acceptance,
- spvmod() and sgvmod() invoke respectively the speaker and signature verification systems,
- spv_out_2 and sgv_out_2 refers to the output of systems for the second attempt.

In the case the outputs of the two systems disagree, the user is called for a second attempt with the trait has been rejected in the first attempt. For the final decision the majority vote approach has been used with the output of the two systems in the first attempt and the output of the system in the third attempt. Table 3 summarizes final results after the second attempt.

Table 3. Experimental Results – 2nd attempt

	AND – Majority
Type I Error Rate (FR)	5,01
Type II Error Rate (FA)	2,14

7   Conclusion

In this paper a multi-modal biometric system for personal verification has been presented and its potentialities have been explored. The system combines the verification decisions obtained through a signature verification module and a speaker verification module. Decision combination is performed at abstract-level, by a simple AND rule with the possibility of a second attempt. The experimental results demonstrate the viability of the proposed approach and lead to further progress in this research area. In fact, it is worth noting that signature and speech are biometric traits which are simple to acquire also on mobile computers and well accepted by users. Therefore, they can be considered valuable traits for internet-based applications, in which personal verification is strongly required.

Of course, in order to achieve better performance, system improvements are necessary, for instance, feature set personalization and the use of more sophisticated combination approaches, like those based on decision combination techniques at measurement-level.

References

1. Urèche, O., Plamondon, R.: Document Transport, Transfer, and Exchange, Security and Commercial Aspects. In: Proc. ICDAR 1999, Bangalore, India, pp. 585–588 (September 1999)
2. Zhang, D., Campbell, J.P., Maltoni, D., Bolle, R.M.(eds.): Special Issue on Biometric Systems. IEEE Trans. on Syst., Man and Cybernetics – Part C 35(3) (August 2005)
3. Boyer, K.W., Govindaraju, V., Ratha, N.K.(eds.): Special Issue on Recent Advances in Biometric Systems. IEEE Trans. on Syst., Man and Cyb. – Part B 37(5) (October 2007)
4. Impedovo, D., Pirlo, G.: Automatic Signature Verification – The State of the Art. IEEE Trans. on Syst., Man and Cybernetics – Part C (September 2008) (in press)
5. Plamondon, R., Lorette, G.: Automatic Signature Verification and Writer Identification – The State of the Art. Pattern Recognition 22(2), 107–131 (1989)
6. Leclerc, F., Plamondon, R.: Automatic Signature Verification: The State of the Art – 1989 1993. International Journal of Pattern Recognition and Artificial Intelligence (IJPRAI) 8(3), 643–660 (1994); In: Plamondon, R.(ed.) Special Issue on Progress in Automatic Signature Verification, Series, MPAI, p. 320. World Scientific (1994)
7. Nalwa, V.S.: Automatic On-line Signature Verification. Proceedings of the IEEE 85(2), 215–239 (1997)
8. Fairhurst, M.C.: Signature Verification Revisited: Promoting Practical Exploitation of Biometric Technology. IEE Electronics and Communication Engineering Journal (ECEJ) 9(6), 273–280 (1997)
9. Plamondon, R., Srihari, S.N.: On line and Off line Handwriting Recognition: A Comprehensive Survey. IEEE Transactions on Pattern Analysis and Machine Intelligence (T-PAMI) 22(1), 63–84 (2000)
10. Vielhauer, C.: A Behavioural Biometrics. Public Service Review: European Union (9), 113–115 (2005)
11. Dimauro, G., Impedovo, S., Lucchese, M.G., Modugno, R., Pirlo, G.: Recent Advancements in Automatic Signature Verification. In: 9th International Workshop on Frontiers in Handwriting Recognition (IWFHR-9), Kichijoji, October 25-29, pp. 179–184 (2004)
12. Campbell, J.P.: Speaker Recognition: A tutorial. Proc. of IEEE, 1437–1462 (1997)
13. Reynolds, D.A.: Speaker Identification and Verification using Gaussian Mixture Speaker Models. Speech Communication, 91–108 (1995)
14. Linares, L.R., Mateo, C.G., Castro, J.L.A.: On combining classifiers for speaker authentication. Pattern Recognition, 347–359 (2003)

15. Dimauro, G., Impedovo, S., Pirlo, G.: A new methodology for the measurement of local stability in dynamic signatures. In: Proc. International Workshop on Frontiers in Handwriting Recognition IV, Taipei, Taiwan, December 7-9, pp. 135–144 (1994)

16. Bajaj, R., Chaudhury, S.: Signature Verification Using Multiple Neural Classifiers. Pattern Recognition 30(1), 1–7 (1997)

17. Ramesh, V.E., Narasimha Murty, M.: Off-line signature verification using genetically optimized weighted features. Pattern Recognition 32(2), 217–233 (1999)

18. Rabiner, L.R., Levinson, S.E.: Isolated and connected word recognition. Theory and Selected applications. IEEE Transactions on Communications 29(5), 621–659 (1981)

19. Di Lecce, V., Dimauro, G., Guerriero, A., Impedovo, S., Pirlo, G., Salzo, A., Sarcinella, L.: Selection of Reference Signatures for Automatic Signature Verification. In: Proc. Of Int. Conference on Document Analysis and Recognition 1999, Bangalore, pp. 597–600 (1999)

20. Dimauro, G., Impedovo, S., Pirlo, G.: On-line Signature Verification through a Dynamical Segmentation Technique. In: Proceedings 3th International Workshop on Frontiers in Handwriting Recognition, NY, USA, pp. 262–271 (1993)

21. Yoshimura, I., Yoshimura, M.: On-line signature verification incorporating the direction pen movement - An experimental examination of the effectiveness. In: Impedovo, S., Simon, J.C. (eds.) From Pixels to Features III - Frontiers in Handwriting Recognition, pp. 353–362. Elsevier Publ., Amsterdam (1992)

22. Rabiner, L.R., Schafer, R.: Digital Processing of Speech Signals, ISBN: 0132136031

23. Hoppenheim, A.V., Schafer, R.W.: Homomorphic Analysis of Speech. IEEE Transaction On Audio and Electroacustics AU-16(2), 221–226

24. Deller, J., Hansen, J., Proakis, J.: Discrete-Time Processing of Speech Signals. Classic Reissue. IEEE Press, Los Alamitos (1999)

25. Impedovo, D., Refice, M.: Modular Engineering Prototyping Plan for Speech Recognition in a Visual Object Oriented Environment. In: Information Science and Applications, pp. 2228–2234 (2005)

26. Impedovo, D., Refice, M.: The Influence of Frame Length on Speaker Identification Performance. In: IEEE Proceedings of IAS (2007)

27. Baum, L., Petrie, T., Soules, G., Weiss, N.: A maximization technique occurring in the statistical analysis of probabilistic functions of Markov chains. Ann. Math. Stat. 14, 164–171 (1970)

28. Dempster, A.P., Laird, N.M., Rubin, D.B.: Maximum likelihood from incomplete data via the EM algorithm. J. Royal Statistical Society 39(1), 1–38 (1977)

29. Rabiner, L.R.: A tutorial on hidden Markov models and selected applications in speech recognition. Proc. of the IEEE 77(2), 257–286 (1989)

30. Suen, C.Y., Nadal, C., Legault, R., Mai, T.A., Lam, L.: Computer Recognition of unconstrained handwritten numerals. Proc. IEEE 80, 1162–1180 (1992)

31. Suen, C.Y., Nadal, C., Legault, R., Mai, T.A., Lam, L.: Computer Recognition of unconstrained handwritten numerals. Proc. IEEE 80, 1162–1180 (1992)

32. Velek, O., Jaeger, S., Nagakawa, M.: A Warping Technique for Normalizing Likelihood of Multiple Classifiers and its Effectiveness in Combined On-line/Off-line Japanese Character Recognition. In: Proc. of IWFHR, Niagara on the Lake (Canada), August 6-8, pp. 177–182 (2002)

33. Ross, A., Jain, A.K.: Multimodal Biometrics: An Overview. In: Proc. 12th European Signal Processing Conference (EUSIPCO), September, Vien, pp. 1221–1224 (2004)

34. Kittler, J., Hatef, M., Duin, R.P.W., Matias, J.: On combining classifiers. IEEE Trans. on Pattern Analysis Machine Intelligence 20(3), 226–239 (1998)

Signature Verification Using a Bayesian Approach

Sargur N. Srihari, Kamal Kuzhinjedathu, Harish Srinivasan,
Chen Huang, and Danjun Pu

Center of Excellence for Document Analysis and Recognition
Department of Computer Science and Engineering
University at Buffalo, State University of New York
{srihari,krk32,hs32,chuang5}@cedar.buffalo.edu

Abstract. The fully Bayesian approach has been shown to be powerful in machine learning. This paper describes signature verification using a non-parametric Bayesian approach. Given sample(s) of Genuine signatures of an individual, the task of signature verification is a problem of classifying a questioned signature as Genuine or Forgery. The verification problem is a two step approach - (i)Enrollment: Genuine signature samples of an individual are provided. The method presented here maps from features space to distance space by comparing all the available genuine signature samples amongst themselves to obtain a distribution in distance space - "within person distribution". This distribution captures the variation and similarities that exist within a particular person's signature; (ii)Classification: The questioned signature to be classified, is then compared to each of the genuine signatures to obtain another distribution in distance space - "Questioned vs Known distribution". The two distributions are then compared using a new Bayesian similarity measure to test whether the samples in the distribution are from the same distribution(Genuine) or not(Forgery). The approach yields improved performance over other non-parametric non Bayesian approaches.

Keywords: Signature verification, Bayesian approach.

1 Introduction

The task of forensic signature verification is to verify if a questioned signature sample is genuine or forgery, when given multiple samples of genuine signatures of an individual. It still remains a challenging problem. Some of the previous research in this field such as [1] have focused on identifying random forgeries but much less has been done for skilled forgeries. Parametric approaches such as Naive Bayes classifiers have been used in [2] but they relied on having signature forgery samples as well, for learning. Also, these methods did not learn from samples specific to an individual. Another important note to consider is that, though multiple genuine signature samples of an individual are available, the number of such available samples for an individual is often very few and hence parameters

S.N. Srihari and K. Franke (Eds.): IWCF 2008, LNCS 5158, pp. 192–203, 2008.
© Springer-Verlag Berlin Heidelberg 2008

estimated for parametric models are bound to be incorrect and noisy. Hence it is intuitive to consider non-parametric approaches for learning a writer specific model. This paper discusses a distance based non-parametric Bayesian approach that attempts to capture the variation and similarities(distance) amongst the genuine signature samples and use it to classify a new questioned sample as genuine or forgery.

2   Feature Extraction and Similarity Measures

The *features* for a given set of signatures of a particular writer can be termed as a set of elements that help uniquely label the samples as belonging to that writer. In this paper, the Gradient, Structural and Concavity (GSC) features are used and they are based on the philosophy that feature sets can be designed to extract certain types of information from the image[3,4]. Gradient features use the stroke shapes on a small scale, structural features use stroke trajectories on an intermediate scale, and concavity features use stroke relationships at long distances. They have been defined and used extensively in[5,6,7,8,9] and a review of the features is in [10]. Figure 1(a) shows an example of a signature, from which a 1024 length binary feature vector have been extracted, as shown in Figure 1(b). The features are now considered representative characteristics of the signature.

In order to compare two samples and to quantify their similarity, a similarity (or distance) measure is used to compute a score that signifies the strength of match between the features of the two samples. It converts the data from feature space to *distance* space. This similarity measure again can be different depending upon the kind of features used. Several similarity measures that can be used with binary vectors have been discussed in [11], in which the *"correlation"* measure

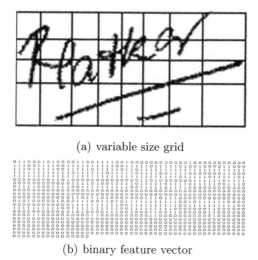

(a) variable size grid

(b) binary feature vector

Fig. 1. Feature computation for signature: (a) variable size grid, and (b) binary feature vector

yields the best accuracy in matching handwriting shapes. It is defined as follows. Let S_{ij} $(i, j \in \{0, 1\})$ be the number of occurrences of matches with i in the first vector and j in the second vector at the corresponding positions, the dissimilarity D between the two feature vectors X and Y is given as follows,

$$D(X, Y) = \frac{1}{2} - \frac{S_{11}S_{00} - S_{10}S_{01}}{2\sqrt{(S_{10} + S_{11})(S_{01} + S_{00})(S_{11} + S_{01})(S_{00} + S_{10})}} \qquad (1)$$

It can be observed that the range of $D(X, Y)$ has been normalized to $[0, 1]$. That is, when $X = Y$, $D(X, Y) = 0$, and when they are completely different, $D(X, Y) = 1$.

3 Verification Model

A writer specific model using a distance based non-parametric approach is explained below. When multiple genuine samples of a writer are available, it is intuitive to learn collectively from all of those samples specific to the writer. The writer specific learning focuses specifically on the writer whose identity needs to be learnt from multiple known samples available and then answers the question if an anonymous questioned sample belongs to this person or not. First, pairs of genuine samples are compared using a similarity measure to obtain a distribution over distances between features of samples – this represents the distribution of the variation/similarities amongst genuine samples – for the particular writer. The corresponding inference method involves comparing the questioned sample against all available genuine samples to obtain another distribution in distance space. Lastly, the two distributions are compared to make a decision of genuine or forgery. These three different steps are explained below.

3.1 Within-Writer Distribution

When there are multiple known samples from a person, it is more intuitive to use all of that information to learn the similarities within and variation across that writer's signature samples, and then use this information as a whole to match against any anonymous sample to test whether the sample belongs to this writer(genuine) or not(forgery).

If a given person has N samples, $\binom{N}{r}$ defined as $\frac{N!}{r!(N-r)!}$ pairs of samples can be compared as shown in Figure 2. In each comparison, the distance between the features is computed. The result of all $\binom{N}{2}$ comparisons is a $\{\binom{N}{2} \times 1\}$ distance vector. This vector is the distribution in distance space for a given writer. A key advantage of mapping from feature space to distance space is that the number of data points in the distribution is $\binom{N}{2}$ as compared to N for a distribution in feature space alone. Also the calculation of the distance between every pair of samples gives a measure of the variation in samples for that writer. In essence the distribution in distance space for a given known writer captures the similarities and variation amongst the samples for that person. Let N be the total number

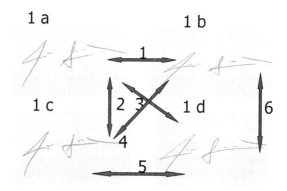

Fig. 2. Samples from one writer provide $\binom{4}{2} = 6$ comparisons

of samples and $N_{WD} = \binom{N}{2}$ be the total number of comparisons that can be made . The within-writer distribution can be written as

$$\boldsymbol{D_W} = (d_1, d_2, \ldots, d_{N_{WD}})^\top \tag{2}$$

where $\top$ denotes the transpose operation.

3.2   Questioned vs Genuine Distribution

Analogous to what we did in Section 3.1 to obtain within-writer distribution, in order to make a decision of genuine or forgery on an unknown signature sample, the questioned sample is compared with every one of the known N samples in a similar way to obtain the Questioned vs Genuine distribution, which is given as follows

$$\boldsymbol{D_{QK}} = (d_1, d_2, \ldots, d_N)^\top \ , \tag{3}$$

3.3   Comparing Distributions

Once the two distributions are obtained, namely the within-writer distribution, denoted D_w (Section 3.1, equation 2), and the Questioned vs Genuine distribution, D_{QK} (Section 3.2, equation 3), the task now is to compare the two distributions to obtain a probability of similarity. The intuition is that if the questioned sample did indeed belong to the ensemble of the knowns, then the two distributions must be the same (to within some sampling noise). There are various ways of comparing two distributions, some of which include Kolmogorov-Smirnov test, Kullback-Leibler divergence and Jensen-Shannon test. These and more have been discussed in detail in [12]. Below we introduce the Bayesian approach that performs better than those discussed in [12]. Since the comparison of distributions is a key step in the inference task, they are explained in detailed in a separate section 4. Before we move onto that section, we want to mention that, the input to the general model should be uni-modal distributed data. In case multi-modal distributions exist, they can be detected by formulating it as a graph partitioning problem. Also, in noisy cases, there are more chances of multi-modal distributions and there the same idea will be helpful.

4   Bayesian Approach to Compare Two Distributions

We refer to the two distributions that needs to be compared as multisets. Given two multisets, the task is that of quantifying how similar they are distributed. The problem is especially difficult for very small multisets (< 20). The solution follows from a random experiment assumed to have given rise to the two distributions. The idea behind the definition of the measure is that if two multisets came from the same distribution, there would be a large number of distributions under which their joint probability would be high. On the other hand if they came from different distributions it is less likely that they are similarly distributed. The number of distributions under which they would jointly have a high probability would also be small. Therefore the average joint probability (or the probability density in the continuous case) of the two distributions under a family of distributions seems to be a good similarity measure.

4.1   Definition

Let S_1 and S_2 be 2 multisets of real numbers of size n and m respectively. To define a similarity measure between S_1 and S_2, we first define a random experiment that produces S_1 and S_2 in the following manner. Let F be a family of probability distributions.

- Two probability distributions D_1 and D_2 are drawn independently from a uniform distribution over F.
- S_1 is obtained by drawing n samples from D_1
- S_2 is obtained by drawing m samples from D_2

We now define the similarity measure to be **the probability that S_1 and S_2 came from the same distribution.** In other words, it is the probability of D_1 and D_2 being the same distribution given S_1 and S_2. Formally, it is the probability $P_F(D_1 = D_2|S_1, S_2)$.

4.2   General Form of the Similarity Measure

We can obtain the expression for the $P_F(D_1 = D_2|S_1, S_2)$ when D_1 and D_2 belong to some family of distributions F as follows.

Note: Since we are considering continuous distributions here, $P_F(D_1 = D_2|S_1, S_2)$ would be the probability density and not the probability. It can therefore take any non-negative real value.

Notation: f, f_1, f_2, f_i denote probability distributions that belong to family F. By marginalizing over F, we obtain

$$P_F(D_1 = D_2 | S_1, S_2) = \int_F P(D_1 = D_2 = f | S_1, S_2) df. \tag{4}$$

By Bayes rule, we have

$$P(D_1 = D_2 = f | S_1, S_2) = \frac{P(S_1, S_2 | D_1 = D_2 = f) \times P(D_1 = D_2 = f)}{P_F(S_1, S_2)} \tag{5}$$

The expression for $P_F(S_1, S_2)$ can be obtained by marginalizing over the values of D_1 and D_2.

$$P_F(S_1, S_2) = \int_F \int_F P(S_1, S_2 | D_1 = f_1, D_2 = f_2) df_1 df_2 \tag{6}$$

Since S_1 and S_2 are independent, this reduces to

$$P_F(S_1, S_2) = \int_F P(S_1 | D_1 = f_1) df_1 \times \int_F P(S_2 | D_2 = f_2) df_2 \tag{7}$$

Using 7 in 5 we have

$$P(D_1 = D_2 = f | S_1, S_2) = \frac{P(S_1, S_2 | D_1 = D_2 = f) \times P(D_1 = D_2 = f)}{\int_F P(S_1 | D_1 = f_1) df_1 \times \int_F P(S_2 | D_2 = f_2) df_2} \tag{8}$$

Using 8 in 4 we have

$$P_F(D_1 = D_2 | S_1, S_2) = \int_F \frac{P(S_1, S_2 | D_1 = D_2 = f) \times P(D_1 = D_2 = f)}{\int_F P(S_1 | D_1 = f_1) df_1 \times \int_F P(S_2 | D_2 = f_2) df_2} df. \tag{9}$$

Since D_1 and D_2 are drawn from a uniform distribution over F, the above expression can be written as

$$P_F(D_1 = D_2 | S_1, S_2) = \frac{Q_F(S_1 \cup S_2)}{Q_F(S_1) \times Q_F(S_2)} \tag{10}$$

where $Q_F(S)$ stands for the marginalized joint probability of the sample multiset S under the family F.

$$Q_F(S) = \int_F P(S | f) df \tag{11}$$

4.3 Expression under the Gaussian Assumption

When F is the family of Gaussian distributions, it is possible to obtain a closed form for Q_F. Sampling or other approximation techniques would have to be resorted to when it is not possible to evaluate the integral in 11.

Equation 11 assumes the following form when F is the family of Gaussian distributions.

$$Q_G(S) = \int_{-\infty}^{\infty} \int_{0}^{\infty} P(S|\mu, \sigma) d\sigma d\mu \tag{12}$$

The following form for $Q_G(S)$ is obtained by evaluating the above integral

$$Q_G(S) = 2^{-\frac{3}{2}} \times \Pi^{\frac{1-n}{2}} \times n^{\frac{-1}{2}} \times C^{(1-\frac{n}{2})} \times \Gamma(\frac{n}{2} - 1) \tag{13}$$

where n is the cardinality of the multiset S and C is given by

$$C = \sum_{x \in S} x^2 - \frac{(\sum_{x \in S} x)^2}{n} \tag{14}$$

Therefore, the distribution similarity measure for two small multisets S_1 and S_2 assuming that they have been drawn from Gaussian distributions is

$$\frac{Q_G(S_1 \cup S_2)}{Q_G(S_1) \times Q_G(S_2)} \tag{15}$$

5   Experiments and Results

5.1   Datasets Description

Two datasets were used in the experiment. The first one is the same dataset used in [13], which contains 55 individuals, each with 24 genuine signatures and 24 forgeries forged by 3 other writers. One example of each of 55 genuine are shown in Figure 3.

Ten examples of genuine of one person (writer No.21) and ten forgeries of that writer are shown in Figure 4. Each signature was scanned at 300 dpi gray-scale

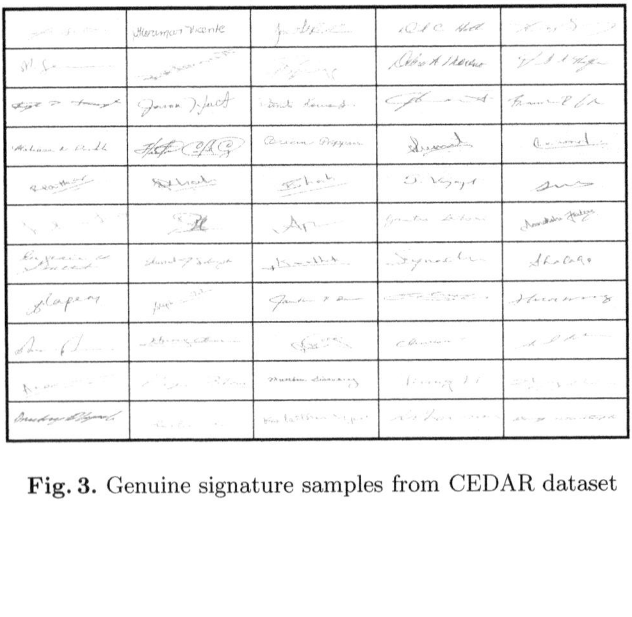

Fig. 3. Genuine signature samples from CEDAR dataset

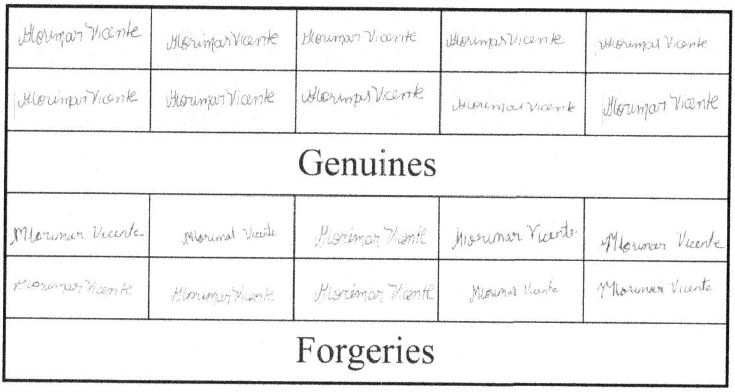

Fig. 4. Samples for writer No.21

and binarized using a gray-scale histogram. Salt pepper noise removal and slant normalization were two steps involved in image preprocessing.

The second dataset used in this experiment was originally created by the American Board of Forensic Document Examiners (ABFDE), and the purpose of it is for academic research and FDE professional development and training. It contains 1564 genuine and forgery signatures of two specimens: "John Moll" and "Marie Wood". Both genuine and forgery have two sub-categories, for genuine, there are normal and disguised signatures; and for forgery, there are spurious and simulated signatures. Spurious signatures are naturally written signatures written by one writer in the name of another writer; no attempt to disguise, distort or at simulating the writing of another. Table 1 shows the number of signatures for each category. For simulated signatures, "John Moll" and "Marie Wood" were simulated by 9 and 13 individuals respectively. In addition, genuine signatures were written in 7 continuous days while simulated samples were created in 5 continuous days. All the signature samples were scanned at 300 dpi gray-scale. One sample image for each category is shown in Figure 5.

5.2   Experiment Setup and Results

Since the two-class classification is based on a similarity (or distance) measure between two distributions, one operating point needs to be determined to

Table 1. Number of signatures for each category

	John Moll	Marie Wood
Normal	140	179
Disguised	35	21
Spurious	90	0
Simulated	449	650
Total	714	850

	John Moll	Marie Wood
Genuine		
Disguised		
Spurious		N/A
Simulated		

Fig. 5. Sample signature images for each category

distinguish between genuine and forgery signatures. In this experiment, this operating point was learned from a training set by using the Bayesian optimal decision rule that minimizes the overall error rate.

In the first dataset, for each test case a writer (amongst a total 55) was chosen and N genuine samples chosen at random of that writer's signature were used for learning. Also the same number of forged signatures of this writer were used for training. The remaining genuine and forgery samples were used for testing. Since the N genuine samples are chosen at random, 5 iterations of test runs were performed (cross-validation approach), such that on each run a different set of N genuine signatures will be used for learning and the remaining $24 - N$ for testing. This also increases the number of test signatures.

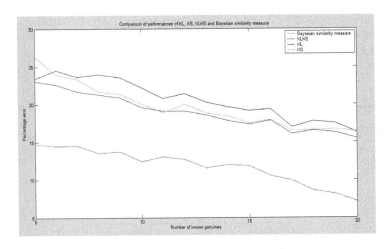

Fig. 6. Comparison of performance of KL, KS, KLKS and the Bayesian similarity measure for different number of knowns genuine signatures

The error rates of the Bayesian approach is compared to other methods such as Kolmogorov-Smirnov test(KS), Kullback Leibler divergence(KL) and also a combined measure(KLKS). These as well as other information theoretic measures and their performance on signature verification have been discussed in detail in [12]. Figure 6 shows the comparison of the performances of KL,KS, KLKS and the Bayesian similarity measure. It can be observed that the Bayesian method outperformed all the other 3 methods about 8% in term of overall error rate.

In the second dataset, for each writer, 100 genuine and forgery samples chosen at random were used for training. The remaining samples were used for testing. Among the 100 genuine samples, N of them chosen randomly were used as the "Known" signatures at the enrollment step. Since the N genuine samples are chosen at random, 10 iterations of test runs were performed (cross-validation approach), such that on each run a different set of N genuine signatures will be used for learning.

Table 2 shows the accuracy for each sample set. In this table, "Genuine" includes only normal signatures, while "forgery" includes both spurious and simulated samples.

Table 2. Bayesian approach on ABFDE dataset

	Genuine(%)	Forgery(%)	Overall(%)
John Moll	93.0	92.2	92.6
Marie Wood	86.1	81.4	83.8
Average	89.6	86.8	88.2

Figure 7 shows the error rates of the Bayesian approach and the best Information Theoretic method, which is the combination measure of Kolmogorov-Smirnov test(KS) and Kullback Leibler divergence(KL) [12].

Testing on disguised signatures is reported separately in Table 3.

Table 3. The performance on disguised signatures

	Bayesian(%)	Information Theoretic(%)
John Moll	59.3	61.1
Marie Wood	69.5	71.9
Average	64.5	66.5

It can be observed that for the second dataset, the performance of Bayesian and Information Theoretic methods are similar. Also, results from both datasets showed that, in general, as the number of "Known" samples is increased, the accuracy will also be increased.

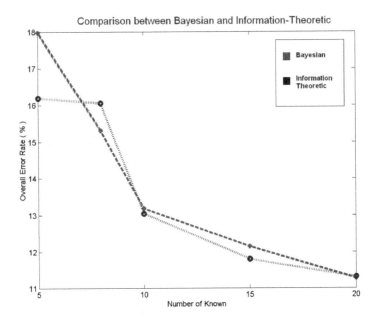

Fig. 7. Overall error rate for both Bayesian and Information Theoretic methods

6   Conclusions

We have discussed a non-parametric distance Bayesian approach for forensic signature verification. The learning strategy involve mapping from feature space to distance space, to learn the variation and similarities that exist amongst the samples for an individual. As the number of samples used for training is increased, the error rate decreases. Experiments show that, while getting similar performance on one dataset, Bayesian approach performs better than other methods by approximately 8% in total error rate on the other dataset.

The domain of signatures and the related experiments performed is just one example of how the approach and the classification techniques can be used. It suffices to say that the Bayesian similarity measure is a general measure that can be used to compare any two multi-sets.

References

1. Sabourin, R., Genest, G., Preteux, F.J.: Off-line signature verification by local granulometric size distributions. IEEE Transactions on Pattern Analysis and Machine Intelligence 19(9), 976–988 (1997)
2. Srihari, S.N., Xu, A., Kalera, M.K.: Learning strategies and classification methods for off-line signature verification. In: Proc.of the 7th Int. Workshop on Frontiers in handwriting recognition(IWFHR), pp. 161–166 (2004)

3. Favata, J., Srikantan, G.: A multiple feature/resolution approach to handprinted digit and character recognition. International Journal of Imaging Systems and Technology 7, 304–311 (1996)

4. Srikantan, G., Lam, S.W., Srihari, S.N.: Gradient based contour encoding for character recognition. Pattern Recognition 29(7), 1147–1160 (1996)

5. Fang, B., Leung, C.H., Tang, Y.Y., Tse, K.W., Kwok, P.C.K., Wong, Y.K.: Off-line signature verification by the tracking of feature and stroke positions. Pattern recognition 36, 91–101 (2003)

6. Lee, S., Pan, J.C.: Off-line tracking and representation of signatures. IEEE Transactions on Systems,Man and Cybernetics 22, 755–771 (1992)

7. Sabourin, R., Plamondon, R.: Preprocessing of handwritten signatures from image gradient analysis. In: Proc.of the 8th Int.Conference on Pattern Recognition, pp. 576–579 (1986)

8. Lin, C.C., Chellappa, R.: Classification of partial 2-d shapes using fourier descriptors. IEEE Transactions on Pattern Analysis and Machine Learning Intelligence, 690–696 (1997)

9. Ammar, M., Yoshido, Y., Fukumura, T.: A new effective approach of off-line verification of signatures by using pressure features. In: Proc.of the 8th Int.Conference on Pattern Recognition, pp. 566–569 (1986)

10. Cha, S.H., Tappert, C.C., Srihari, S.N.: Optimizing binary feature vector similarity measure using genetic algorithm and handwritten character recognition. In: Proceedings of the Seventh International Conference on Document Analysis and Recognition, vol. 2, p. 662 (2003)

11. Zhang, B., Srihari, S.: Properties of binary vector dissimilarity measures, Cary, North Carolina (September 2003)

12. Harish Srinivasan, M.B., Srihari, S.: Machine learning for person identification. In: SPIE Conference on Homeland Security, Orlando, FL, pp. 574–586 (March 2005)

13. Kalera, M.K., Zhang, B., Sriahri, S.N.: Off-line signature verification and identification using distance statistics. In: Proc.of the Int.Graphonomics society Conference, pp. 228–232 (2003)

Stroke-Morphology Analysis Using Super-Imposed Writing Movements

Katrin Franke

Norwegian Information Security Laboratory, Gjøvik University College, Norway
kyfranke@ieee.org

Abstract. Handwritten signatures play an important role in daily life. Consequently, there is a strong need for objective signature evaluation. This paper focuses on a new computational method for discovering and evaluating ink-trace characteristics related to the writing process. It aims (i) to provide a scientific basis for procedures applied in forensic casework and (ii) to derive advanced computational methods for the analysis of signature-stroke morphology. It work towards methods for inferring writer-specific behaviors from the residual ink trace. The respective micro-patterns, caused by biomechanical writing and physical ink-deposition processes, provide important clues for the analysis. These inner ink-trace characteristics of signatures, which are determined by the individual movements of a person, will be studied in depth, taking into account the effects of writing materials, such as the type of pen used. By means of recorded and super-imposed writing movements, ink traces are sampled, and local ink-trace characteristics are encoded in one feature vector per sample record. These data establish a sequence which faithfully reflects the spatial distribution of ink-trace characteristics and solves problems of methods previously available.

Keywords: residual ink trace, writing behavior, signature verification, combined on/offline handwriting analysis, forensic investigation.

1 Introduction

A forensic expert applies various techniques for the examination of graphical signature characteristics as well as for the examination of the physical and chemical properties of the writing materials. However, the visual inspection of the signature pattern only renders subjective and descriptive results and is therefore subject to frequent criticism in courts of justice. A major drawback has hitherto consisted in the lack of a scientific basis for deducing the particularities of the writing process from the residual ink trace.

Our conducted technology research works towards a multi-disciplinary approach to signature analysis, under consideration of perspectives and findings from a number of disciplines, including forensics, human-movement science, image processing and pattern recognition as well as robotics. The research method applied is Analysis by Synthesis. Behavioral characteristics of the signing process and stroke morphology of ink traces are investigated. A major challenge

S.N. Srihari and K. Franke (Eds.): IWCF 2008, LNCS 5158, pp. 204–217, 2008.

in this context is the requirement to explore effects on the ink trace under strictly controlled conditions. Thus, the experimental facilities for studying the ink-deposition process was established in advance. On the basis of theoretical considerations and empirical studies, computational methods are designed for signature preprocessing, feature extraction and analysis.

The approach presented in this paper is inspired by forensic expertise and elaborated by digital signal processing and pattern recognition. In accordance with an *Ink-Deposition Model* [1], which incorporates the physical ink-deposition process, handwritten traces are normalized and segmented into regions of relatively similar ink intensity. In addition, a fundamental approach for superimposition, combined analysis and cross-validation of recorded temporal writing / tracing movements and digitized ink traces is established.

The paper is organized in six sections. Related work is briefly discussed in section 2. The method for ink-deposition analysis using recorded writing movements is outlined in Section 3. Core Algorithms are described in the subsequent Section 4. The experimental setup and results are given in Section 5. The final Section 6 provides the conclusions and hints to further research.

2   Related Work

According to forensic expertise, e.g. [2,3], the most reliable method for drawing conclusions regarding the authorship of a questioned signature specimen is to infer writing movements, in particular the applied kinematics and kinetics e.g. [4,5,6,7], from inner ink-trace characteristics, also referred to as stroke morphology. The underlying interaction of biomechanical handwriting and physical ink-deposition processes had hitherto not been studied systematically. A scientifically founded basis is needed in order to preclude criticism and to technologically upgrade the classical visual inspection. To this end, the factors that may influence the ink deposition on paper have to be taken into account.

Various approaches to recover temporal information from static handwriting specimens already exist. Procedures in forensic examination are mainly based on the microscopic inspection of the writing trace and hypotheses regarding the underlying writing process, e.g. [2,3,8]. The techniques applied in the field of image processing and pattern recognition can be divided into (i) mathematical methods for estimating the temporal order of stroke production [9,10,11,12], (ii) methods inspired by motor-control theory for recovering temporal features on the basis of stroke geometries, such as curvature [13,14], and (iii) methods for analyzing stroke thickness and / or stroke-intensity variations [15,16,17,18]. However, these previous works in the field of image processing did not sufficiently take into consideration the physical properties and influences of writing materials like pen and ink. In addition, they did not preserve the spatio-temporal relationship of the deposited ink.

Recorded writing movements can be used for the analysis of ink traces. There are a number of possible analysis procedures that take temporal information into account. Such methods can be subdivided into three primary groups, i.e. (1) the reconstruction of the stroke sequence [19,20], (2) the tracing of the handwritten ink traces [21], and (3) the assignment of temporal handwriting characteristics, e.g. relative pen force and writing velocity to static ink-trace characteristics [22]. Additionally, the origin of the online handwriting data has to be taken into account. The following scenarios can be distinguished: (A) Simultaneously produced on- and offline data, i.e. written with an ink pen on a sheet of paper placed on an electronic tablet, (B) Traced online data entered by a forensic expert by retracing an offline sample, (C) Online samples written by the same person at an earlier / later date, and (D) Online samples written by another person. The combined usage of on- and offline data offers a wide range of analysis and cross-validation methods.

Recorded writing movements can also be used for the replication of writing behaviors [23]. Time series of temporal handwriting data can be translated into machine commands to control a writing robot. Subsequently, the writing robot is able to repeat exactly the same movement multiple times and to synthesize ink traces, which are not influenced by natural human variation. These mechanically produced ink traces can subsequently be used for a dedicated analysis.

3 Method Overview

A new computational method is proposed here for analyzing the stroke morphology of signatures and, in particular, for analyzing the relative amount of ink deposited on paper. It supports the reconstruction of the stroke sequence as well as the sensing of the ink trace by means of superimposed writing movements. For a schematic overview of the approach, especially for preprocessing and feature extraction, see Figure 1a. In detail, the method comprises the following steps:

Data capture (DC). Optical scanning of the paper document carrying the ink trace. The online trajectory is recorded by means of an electronic pen tablet.

Preprocessing module (PP)

- *DPP* Offline document preprocessing for removing backgrounds and imprints using methods described in [24].
- *AS* Alignment / Superposition of on- and offline data for sensing the ink flow (Figure 1b). It has to be taken into account that currently available tablet technologies are partly restricted with regard to their signal fidelity [25], and that ink tracing, e.g. performed by a forensic expert, is not sufficiently accurate, so that the captured pen position cannot be used in a computational method. An automatic procedure for aligning on- and offline data is described in section 4.
- *FSP / RS* Filtering and equidistant re-sampling of the online data in order to provide equally distributed measurement points and to indicate *stroke intersections*, *cross-* and *near points*, and *stop points* (Figure 1b). The challenge of near points in offline analysis was discovered by Doermann et al. [26]. However, online data can help to handle arising difficulties in the offline analysis later on. It must be noted that for the filtering of near and cross points some heuristics about the digitized ink trace and its line width are taken into account. It results

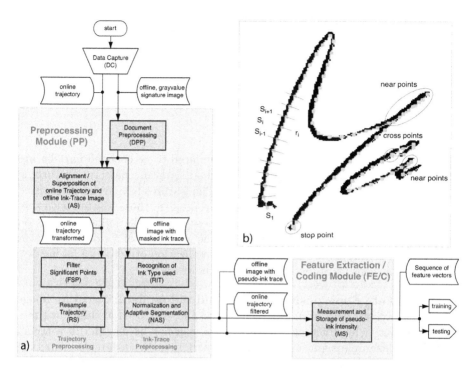

Fig. 1. a) Schematic overview of the preprocessing and feature extraction applied in ink-deposition analysis. The consideration of recorded writing movements demands specific methods for preprocessing the online trajectories. In addition, the offline ink traces have to be normalized. b) Preprocessed ink trace and superimposed online trajectory. Ambiguous stops as well as the cross and near points are marked. In addition, sample points and rulers are illustrated, and the normalized ink deposit is sensed.

in the marking of an extended number of samples of the online trajectory (see Figure 1b).

- *RIT*   Recognition of the ink type used [27] in order to determine whether the stroke-morphology characteristics are suitable for a more detailed analysis that takes the appropriate Ink-Deposition Model [1] into account.

- *NAS*   Normalization and adaptive segmentation of the ink trace in order to quantize ink intensities. In this way the ink trace is converted into pseudo-ink segments that become independent of the particular ink used. For details see the upcoming Section 4.

Feature extraction and coding module (FE/C)

- *MS*   Measurement of normalized ink intensity along the superimposed online trajectory, and storing of the sensed characteristics in a sequence of feature vectors. These methods for feature extraction and encoding into numerical parameters, which represent the relative ink deposit on paper, are covered by Section 4.

Analysis and validation module (AV)

- *VFS* Validation of the sequence of feature vectors obtained from a questioned signature specimen by comparing them with those of a known reference sample. The chosen sequential feature encoding motivates the employment of well-established verification techniques in online signature analysis, e.g. [28,29,30,31,32,33].

4   Core Algorithms

For a computer-based analysis the ink deposits need to be encoded into feature vectors. These must be valid for each individual signature pattern. Disturbing influences need to be eliminated in advance. The features have to be normalized to facilitate the cross-validation of different probes, e.g. those written with specific kinds of pens and ink. The procedures described in the following focus on exactly these requirements, especially on (i) the alignment of residual ink traces and recorded writing movements, (ii) the normalization of ink-trace intensities, and on (iii) the sensing of ink deposit along the trace.

AS Alignment / Superposition of on- and offline data: Signature image and sign-behavioral data from different sources, i.e. produced at different times and / or produced by different human writers, can not be overlaid in a trivial manner. For example signature-ink tracing performed by a forensic expert is not sufficiently accurate. Moreover, available tablet technologies are partly restricted with regard to their signal fidelity [25], causing displacement errors of the recorded pen-position signal. Consequently, scaling, rotating, translating, matching, and non-linear morphing of the online sample position is demanded that facilitates the alignment / superimposing of online pen trajectories and offline ink traces. The procedure described in the following allows for the processing of (i) on- and offline signature data that have been simultaneously produced, and (ii) digitized ink strokes that have been traced with an electronic pen-tablet or mouse. A more elaborated procedures that matches signatures strokes [34] is needed to adjust on- and offline data that were produced at different times, either by the same / genuine writer or an impostor.

Superimposing displaced on- and offline data is performed as follows:

1. Binarize digitized ink-trace image I_{off} with an appropriate document preprocessing method [24].
2. Estimate the average stroke width $\bar{w}_{off}$ of the binarized ink trace.
3. Produce a second binary handwriting image I_{on} in a similar spatial resolution by means of the recorded handwriting trajectory (online signal), with the estimated stroke width of $\bar{w}_{off}$ and an appropriate brushing function [35].
4. For on- and offline signature specimens, compute the centers of gravity $C_{on,off}$ of all black image elements, with $C_{on} = cog\{I_{on}|I_{on}(x,y) = 0\}$ and $C_{off} = cog\{I_{off}|I_{off}(x,y) = 0\}$, respectively.
5. Perform principal component analysis [36] on obtained handwriting images I_{on} and I_{off} separately. By sorting the eigenvectors E in the order of descending eigenvalues (largest first), one can find the directions $\mathbf{D}_{xy,on}$ and $\mathbf{D}_{xy,off}$

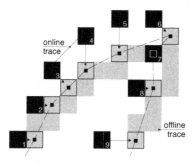

Fig. 2. Schematic overview of the super-positioning of on- and offline specimens. A general alignment of the on- and offline strokes is followed by the morphing of online sample points to the nearest offline trace element.

with the largest variance of the data, with $\boldsymbol{D}_{\text{xy, on}} = E_1\{I_{\text{on}}|I_{\text{on}}(x,y) = 0\}$ and $\boldsymbol{D}_{\text{xy, off}} = E_1\{I_{\text{off}}|I_{\text{off}}(x,y) = 0\}$, respectively.

6. Determine translation $T_{\text{xy, on off}}$ between on- and offline handwriting specimens by means of computed centers of gravities C_{on} and C_{off}.
7. Determine rotation $R_{\text{on-off}}$ by using the derived principal components $\boldsymbol{D}_{\text{xy, on}}$ and $\boldsymbol{D}_{\text{xy, off}}$.
8. Perform affine transformation A_{on} on the online signal using the translation $T_{\text{xy, on off}}$, rotation $R_{\text{on-off}}$ and scaling factor $S_{\text{on:off}}$ according to the ratio between the spatial resolution of the tablet data and the offline image.
9. For all globally aligned, online sample points find the nearest offline image element and translate online data point to this position (compare Figure 2).

NAS Normalization and adaptive segmentation of the ink trace: In order to ensure validity regardless of the particular pen / ink type used, ink traces extracted from document backgrounds need to be normalized. Different approaches can be found in literature, for example, the *Densitron*-approach by Grube [37,38], and threshold techniques applied, e.g., by Sabourin et al. [17] or by Ammar et al. [15]. The latter two are adaptive methods, but unfortunately they do not appropriately take into consideration the characteristics of the ink-intensity distribution. Particularly the distinct characteristics of ink-intensity-frequency plots produced by solid, viscous and fluid inks, are not taken into account (compare [1]). The methods only assume the presence of so-called *high-pressure regions* that correspond to greater pen-tip forces. The high-pressure regions are defined as ink intensities exceeding a threshold P_0. The threshold P_0 is either determined by the well-known Otsu method [39], by Ammars α-cut procedure [15] or empirically preset, e.g. [40,21,41]. In contrast, the previously mentioned *Densitron*-approach is not adaptive. It only represents an intensity range, e.g. 64 intensity levels, by one pseudo-color. Since these intensity ranges were predefined, and not adaptively adjusted to the support, the mean or median intensity of the ink-intensity distribution under investigation, traces by different pens / inks could not be cross-validated. Nevertheless, the basic idea of the

Densitron-approach could support the modeling of the entire ink trace. To allow for a normalization and validity irrespective of the particular pen / ink used, the approach was further elaborated by the author [42]. In addition, a pilot study was conducted and the newly derived *adaptive Densitron* procedure was compared with an alternative approach as well as with the method proposed by Ammar et al. [15]. Further discussions on the Otsu method [39] are unnecessary, since the limitations of this method in handling heavily skewed distributions have already been addressed in [1].

For a formal description of the methods investigated in the pilot study, let us first define a digital image $I(x,y)$ with $0 \leq x < X$ and $0 \leq y < Y$. The background was removed in order to filter a set T of image coordinates $(T \subset I(x,y))$ that belong to a digitized ink trace. For the normalization of ink-trace intensities $I(x,y)$ a transformation into so-called pseudo-ink intensities $\tilde{I}(x,y)$ was performed. Afterwards, for each T the intensity-frequency plot was computed by

$$h(k) = \#\{(x,y)|I(x,y) = k; (x,y) \in T\} \tag{1}$$

with $0 \leq k \leq g_{\max}$. The derived ink-intensity distribution provides the basis for the upcoming normalization. Specific procedures in accordance with *(NAS 1)* Ammar's *α-cut procedure*, *(NAS 2)* the *adaptive Densitron with equal range* and *(NAS 3)* the *adaptive Densitron with equal area* are detailed in the following.

Ad. NAS 1 – α-cut procedure: The approach proposed by Ammar [15] is defined as:

$$H(k) = \frac{h(k)}{\max_k h(k)} \tag{2}$$

$$P_0 = \max_k \left(\{k \mid H(k) \geq \alpha_{\text{cut}}\}\right) \tag{3}$$

Studies on various ink distributions have revealed that dubious segments occur especially in the case of writers who "glide" across the paper, since the intensity distribution is shifted to the right and therefore almost the entire ink trace is falsely interpreted as high-pressure region.

Ad. NAS 2 – Adaptive Densitron with equal range: Our first implementation of an adaptive Densitron was directly inspired by the original approach; the support of the distribution was segmented into a number N of ranges of equal size:

$$\Delta s = \frac{\max_k(\,\text{support}(h(k))\,) - \min_k(\,\text{support}(h(k))\,)}{N} \quad \text{with} \tag{4}$$

$$\text{support}(h(k)) = \{k \mid h(k) > 0, \quad k = 0, \ldots, g_{\max}\} \tag{5}$$

$$s_0 = \min_k(\,\text{support}(h(k))\,) \quad \text{and} \tag{6}$$

$$s_i = \{k \mid s_0 + (i-1)\Delta s \leq k \leq s_0 + i\Delta s\} \quad \text{with} \quad 1 \leq i \leq N \tag{7}$$

The method, however, does not take the skewness of the distribution into account and yields weak segmentation results for viscous and fluid inks.

Ad. NAS 3 – Adaptive Densitron with equal area: The final version of the adaptive Densitron was designed in such a way that for the given $l_0 = 0$ and $l_N = 254$, l_i segments are chosen by

$$\forall_{i=2,\dots,N} \left(\sum_{k=l_{i-1}}^{l_i} h(k) = \sum_{k=l_{i-2}}^{l_{i-1}} h(k) \right) \tag{8}$$

This approach ensures the consideration of the specific characteristics of ink-intensity distribution and facilitates a reliable ink-trace normalization and transformation into pseudo-ink segments. Consequently, the cross-validation of writing traces produced with different pens / inks becomes feasible.

MS Measurement / Storage of pseudo-ink intensity: In order to support an automatic validation of ink deposits, the pseudo-ink segments have to be sensed and represented by sequences of the numerical feature vectors. Different approaches can be chosen to create these sequences, and therefore pilot tests were carried out to study three of them: *(MA 1)* the *x-projection, (MS 2)* the *vertical scan* and *(MS 3)* the *recovering of stroke sequences.* As a consequence, we decided to use online data for creating the feature sequences.

Ad. MS 1 – X-Projection: According to Ammar et al. [15] all segments of similar pseudo-ink are projected onto the x-axis of the image. The projection profile was scanned in horizontal direction, and thus the frequency of elements per image row determines the x-projection sequence. The sequence length corresponds to the horizontal extension of the signature pattern. This method is highly insensitive to local characteristics of pseudo-ink segments. Particularly the size of segments and their spatial distribution are poorly represented (compare Figure 3a). As observed in the previous empirical studies [43], forgers do not always apply more pen-tip force, and their writing velocity is not necessarily reduced, which would lead to a greater amount of ink deposit. Rather, forged signatures are subject to local variations and adaptation strategies that correlate with the complexity of the motor task. Hence, he / she performs movements with more force impulses and pauses, resulting in small pseudo-ink segments, distributed along the entire writing trace. Applying the x-projection strategy for sensing ink-trace characteristics does not lead to an adequate representation of these specific phenomena, as demonstrated in Figure 3a.

Ad. MS 2 – Vertical scan: In accordance with Sabourin et al. [44], distributed segments are collected by vertical scanning, similar to standard image filtering. The size of each pseudo-segment (amount of image elements) is stored. The number of segments within the image determines the length of the sequence. In contrast to the first approach, each segment is handled separately. However, the location of the segment is not represented adequately, and the approach is not immune to slight local variations. For example, a small shift of a segment can lead to a total disorder of the collected segments (compare Figure 3b). Even a slight modification of the approach by additionally storing the x- and y-coordinate of the center of gravity for each segment, and by subsequently taking

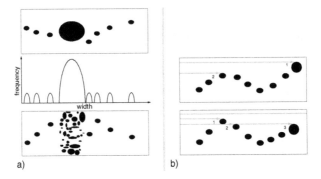

Fig. 3. a) Feature-sequence creation for distributed connected components. The *X-Projection* [15] leads to a loss of spatial distribution, and larger and smaller segments cannot be differentiated, and b) Feature-sequence creation for distributed connected components. *Vertical Scanning* [44] can lead to false representations of the order, if the overall pattern is slightly distorted.

into consideration the spatial distribution of the segments, did not yield any relevant results.

Ad. MS 3 – Recovery of stroke sequences: No doubt, a more natural image-scan path for gathering the pseudo-ink segments would follow the writing trace. A variety of mathematical methods [9,10,11,12,14] for detecting the stroke sequences of handwritten words have been published. So far, attempts to transfer these approaches to signatures have not been successful, since European, and particularly Latin signatures, are frequently very complex, illegible man-made patterns. The underlying heuristics and optimization criteria are not robust with regard to complex trace intersections. Note that even in some difficult cases in forensic casework, human experts are barely able to reestablish the order of the ink traces. To deal with such difficulties Doermann et al. [16,21] proposed a manual tracing of writing traces. This continues to be an adequate solution for a forensic assistance system, but it is inapplicable for automated check processing in banks. In addition, it can be extremely labor-intensive if a large amount of handwritten text is to be analyzed, e.g. in forensic applications where handwriting and signature probes are taken from suspects. Fortunately, modern electronic devices bring some relief. Electronic pens with standard ballpoint refills are now available. These can be employed while taking writing probes in criminal investigations, and they can also be employed for signing a bank check at the counter. In this way simultaneously produced on- and offline data samples become available and can be analyzed in a sophisticated manner [45,21,46].

In order to measure / sense the normalized ink deposit along the writing trace, sample points S_i of the superimposed and spatial-equidistant resampled online trajectory S of length I are employed. The ink sensing is inspired by Doermann [47] who studied ink-intensity profiles orthogonal to the trace direction. Elaborating on this, the direction β_i for a specific measuring line r_i in sample point $S_i\{x_i, y_i\}$ is defined as:

$$\beta_i = \frac{1}{2} \arcsin \sqrt{\frac{(y_{i+1} - y_i)^2 + (x_{i+1} - x_i)^2}{(y_i - y_{i-1})^2 + (x_i - x_{i-1})^2}} \tag{9}$$

whereby $S_{i-1}\{x_{i-1}, y_{i-1}\}$ and $S_{i+1}\{x_{i+1}, y_{i+1}\}$ denote the prior and succeeding sample of S_i, respectively. For the first sample S_1 and last sample S_I in the online trajectory the orthogonal of the related step segment is used. Along the defined 1-pixel-wide ruler r_i, which is a Bressenham line in direction β_i, the frequency of pseudo-ink-trace intensities is determined. The obtained frequency plot $v_i(k)$ per sample S_i is encoded in a feature vector v_i of length N. This vector length N corresponds to the number of pseudo-ink intensities.

$$v_i(k) = \#\{(x, y) | \tilde{I}(x, y) = k; (x, y) \in r_i\} \quad \text{with} \quad 0 \le k < N \tag{10}$$

Subsequently, the ink deposit along the trace is stored as sequence $\{S_i\}$ of feature vectors v_i. The quantization of the original ink-intensity profile ensures (i) validity irrespective of the pen used, (ii) fast / easy numeric analysis and (iii) greater robustness in the upcoming cross-validation of ink traces.

An alternative approach for sensing the ink deposit along an online trajectory needs to be mentioned here. In [21], Guo et al. propose the utilization of pen records obtained by the manual tracing of the digitalized ink strokes. They do not refer to any preprocessing of the online trajectory. The careful consideration of the approach shows that: (i) due to the search of near "dark points" the sampling along the ink trace may not be equally distributed. Specific trace segments can be over- or underrepresented in the final feature sequence. (ii) The threshold for determining "dark points" is a fixed one and may not adequately model the underlying ink-deposit characteristics. In the best of cases it may not be robust, if different ink colors of ballpoint-pen refills, like light blue and black, are used for writing. The computational method proposed in this paper eliminates these drawbacks.

5   Experimental Results

In order to lay the scientific foundations for inferring kinematics and kinetics from the residual ink trace, the following experiment focuses on the stability of ink distributions. As an extension of our experiment performed in [23], it seemed appropriate to investigate to which extent the line quality of signature samples will differ if they are written with exactly the same movements, e.g. by the robot, but by using different pens with the same ink type. The same data set as for stroke-phenomena analysis [23] was used. 26 ballpoint pens were taken to produce 10 signature probes per pen by means of a writing robot. This robot reproduces exactly the same pen trajectory and pen-force time function [43]. The experiments were conducted with $80\,\mathrm{g}\,/\,\mathrm{m}^2$ white copy paper and a soft writing pad, consisting of five of these paper sheets. After the robotic signature synthesis the paper sheets were optically scanned, using a calibrated image scanner with a spatial resolution of $300\,\mathrm{dpi}$ and 8 bit grayvalues. The writing movements of

Table 1. Results of the ink-deposition analysis of signatures produced by the robot with 26 different ballpoint pens and 10 samples each. See Tables 2 and the text for details.

Table 2. Bar-chart of analysis results regarding signature-ink deposits produced by the writing robot with 26 different ballpoint pens and 10 samples per pen (see also Table 1)

	Ink	
	intra-group	inter-group
Median	100.0 %	93.2 %
Quantil 03	100.0 %	88.8 %
Min	30.0 %	16.0 %

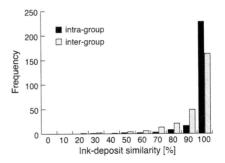

the robot were recorded by an electronic pen-tablet (200 Hz sampling rate and 2540 dpi spatial resolution), thus matching online data were available for superimposing, sensing and analyzing the ink traces. The preprocessed online data were used to sense normalized ink traces, and to generate feature sequences for automatic comparison.

Experiments were conducted to compare the normalized ink deposits of all samples produced with one pen (intra-group), and of all samples produced with the 26 different ballpoint pens (inter-group). The results are listed in Table 1. The ink deposits of samples written with the same pen are highly concurrent and yield a median recognition rate of 100.0 %. For the cross-validation of samples written with different pens, the recognition rate drops slightly to 93.2 % (see Table 1, 2). A closer examination revealed that very often the first trace samples per pen probe produce rather mediocre recognition results. In these cases the ink was not properly extracted from the ink chamber and, as a consequence, less ink was deposited on the paper (ink-free begin strokes). Since specific characteristics of a pen can cloak the (bio)-mechanical effects, it is always advisable to check for defects in the writing instrument. Taking this fact into account, one can conclude that similar writing movements will lead to similar ink deposits on paper, even if different pens of the same ink-type class are used.

6   Conclusions and Future Directions

The new computational method proposed in this section aims at the evaluation of ink-trace characteristics that are affected by the interaction of biomechanical writing and physical ink-deposition processes. The analysis has focused on the ink intensity, which is sensed along the entire writing trace of a signature. This specific analysis was motivated by empirical findings in the field of forensic science, which revealed that mimicked handwriting, which is produced less fluently with many pen-force pulses, will cause disturbances in the inner ink-trace characteristics. In contrast to previous attempts at establishing a computer-based method, the approach presented here introduces new concepts in order to improve the reliability and reproducibility of analysis results. Especially the usage of superimposed, filtered online data makes it possible to take the stroke sequence into consideration. The adaptive segmentation of ink-intensity distributions takes the influences of different writing instruments into account and supports the cross-validation of different pen probes. The analysis is firmly rooted due to the compliance with rules of digitalization and due to sophisticated methods for the removal of document backgrounds and imprints.

From the overall results one can conclude that the proposed computational method is suitable for the analysis of ink deposits along the writing trace. This is also supported by our initial studies on the stability of single stroke phenomena, such as ink drops, striation or feathering [23]. Studies in the forensic field have hitherto been of a descriptive nature. This paper provides all the prerequisites for a systematic analysis of ink deposits under strictly controlled conditions. Along with the experimental results obtained, it lays the methodological foundations for further research, development and forensic casework in the computer-based analysis of signatures. The procedures facilitate elaborate studies on the biomechanics and physical interaction processes, which should especially focus on the validation of samples written by the same human writer with different pens, but also on the cross-comparison of authentic and mimicked handwriting.

References

1. Franke, K., Rose, S.: Ink-deposition model: The relation of writing and ink deposition processes. In: Proc. 9th International Workshop on Frontiers in Handwriting Recognition (IWFHR), Tokyo, Japan, pp. 173–178 (2004)
2. Hecker, M.: Forensische Handschriftenuntersuchung. Kriminalistik (in German) (1993)
3. Michel, L.: Gerichtliche Schriftvergleichung. De Gruyter (in German) (1982)
4. Leung, S., Cheng, Y., Fung, H., Poon, N.: Forgery I: Simulation. Journal of Forensic Sciences 38(2), 402–412 (1993)
5. Leung, S., Cheng, Y., Fung, H., Poon, N.: Forgery II: Tracing. Journal of Forensic Sciences 38(2), 413–424 (1993)
6. Schomaker, L., Plamondon, R.: The relation between pen force and pen point kinematics in handwriting. Biological Cybernetics 63, 277–289 (1990)
7. van Gemmert, A.: The effects of mental load and stress on the dynamics of fine motor tasks. PhD thesis, Catholic University, Nijmegen, The Netherlands (1997)
8. Hilton, O.: Scientific Examination of Questioned Documents (revised edition) edn. CRC Serie in Forensic and Police Science. CRC Press, Boca Raton (1993)
9. Abuhaiba, I., Ahmed, P.: Restoration of temporal information in off-line arabic handwriting. Pattern Recognition 26(7), 1009–1017 (1993)
10. Boccigone, G., Chianese, A., Cordella, L., Marcelli, A.: Recovering dynamic information from static handwriting. Pattern Recognition 26(3), 409–419 (1993)
11. Jäger, S.: Recovering Dynamic Information from Static, Handwritten Word Images. PhD thesis, Albert-Ludwigs-Universität, Freiburg im Breisgrau (1998)

12. Lallican, P.: Reconnaissance de l'Ecriture Manuscrite Hors-ligne: Utilisation de la Chronologie Restaurée du Tracé. PhD thesis, IRESTE, Universite de Nantes, Nantes (1999)
13. Plamondon, R.: The origin of 2/3 power law. In: Proc. 8th Conference of the International Graphonomics Society (IGS), Genova, Italy, pp. 17–18 (1997)
14. Plamondon, R., Privitera, C.: The segmentation of cursive handwriting: An approach based on off-line recovery of the motor-temporal information. IEEE Transactions on Image Processing 8(1), 80–91 (1999)
15. Ammar, M., Yoshida, Y., Fukumura, T.: A new effective approach for off-line verification of signatures by using pressure features. In: Proc. 8th International Conference on Pattern Recognition, Paris, France, pp. 566–569 (1986)
16. Doermann, D., Rosenfeld, A.: Recovery of temporal information from static images of handwriting. International Journal of Computer Vision 15, 143–164 (1995)
17. Sabourin, R., Plamondon, R., Lorette, G.: Off-line identification with handwritten signature images: Survey and perspectives. In: Baird, H., Bunke, H., Yamamoto, K. (eds.) Structured Document Image Analysis, pp. 219–234. Springer, Heidelberg (1992)
18. Wirotius, M., Vincent, N.: Stroke inner strukture invariance in handwriting. In: Teulings, H., van Gemmert, A. (eds.) Proc. 11th Conference of the International Graphonomics Society (IGS), Scottsdale, Arizona, USA (2003)
19. Viard-Gaudin, C., Lallican, P., Knerr, S., Binter, P.: The IRESTE On/Off (IRONOFF) dual handwriting database. In: Proc. International Conference on Document Analysis and Recognition (ICDAR), Bangalore, India, pp. 455–458 (1999)
20. Nel, E., du Preez, J., Herbst, B.: Estimating the pen trajectories of static signutres using hidden markov models. IEEE Transaction on Pattern Analysis and Machine Intelligence (PAMI) 27(11), 1733–1746 (2005)
21. Guo, J., Doermann, D., Rosenfeld, A.: Forgery detection by local correspondence. International Journal of Pattern Recognition and Artificial Intelligence 15(4), 579–641 (2001)
22. Franke, K., Grube, G., Schmidt, C.: Assistance system for the analysis of ink deposit. In: Proc. 4th International Congress of the Gesellschaft für Forensische Schriftenuntersuchung (GFS), Hamburg, Germany (1999)
23. Franke, K., Schomaker, L.: Robotic writing trace synthesis and its application in the study of signature line quality. Journal of Forensic Document Examination 16(3), 119–146 (2004)
24. Franke, K., Köppen, M.: A computer-based system to support forensic studies on handwritten documents. International Journal on Document Analysis and Recognition 3(4), 218–231 (2001)
25. Franke, K.: Capturing reliable data for computer-based forensic handwriting analysis. In: IEEE Three-Rivers Workshop on Soft Computing in Industrial Applications (SMCia), Passau, Germany, pp. 115–120 (2007)
26. Doermann, D., Intrator, N., Rivlin, E., Steinherz, T.: Hidden loop recovery for handwriting recognition. In: Proc. 8th International Workshop on Frontiers in Handwriting Recognition (IWFHR), Niagara-on-the-Lake, Canada, pp. 375–380 (2002)
27. Franke, K., Bünnemeyer, O., Sy, T.: Writer identification using ink texure analysis. In: Proc. 8th International Workshop on Frontiers in Handwriting Recognition (IWFHR), Niagara-on-the-Lake, Canada, pp. 268–273 (2002)
28. Dolfing, H.: Handwriting Recognition and Verification: A Hidden Markov Approach. PhD thesis, Eindhoven University of Technology, Eindhoven, The Netherlands (1998)

29. Gupta, G., McCabe, A.: A review of dynamic handwritten signature verification. Technical report, James Cook University, Townsville, Australia (1997)
30. Leclerc, F., Plamondon, R.: Automatic signature verification: the state of the art 1989-1993. International Journal of Pattern Recognition and Artificial Intelligence 8, 643–660 (1994)
31. Plamondon, R., Lorette, G.: Automatic signature verification and writer identification - the state of the art. Pattern Recognition 22, 107–131 (1989)
32. Schmidt, C.: On-line Unterschriftenanalyse zur Benutzerverifikation. PhD thesis, RWTH Aachen University (in German) (1998)
33. Wirtz, B.: Segmentorientierte Analyse und nichtlineare Auswertung für die dynamische Unterschriftsverifikation. PhD thesis, Technische Universität München (in German) (1998)
34. Franke, K., Zhang, Y., Köppen, M.: Static signature verification employing a Kosko-Neuro-Fuzzy approach. In: Pal, N., Sugeno, M. (eds.) AFSS 2002. LNCS (LNAI), vol. 2275, pp. 185–190. Springer, Heidelberg (2002)
35. Velek, O., Liu, C.L., Nakagawa, M.: Generating realistic kanji character images from on-line patterns. In: Proc. 6th International Conference on Document Analysis and Recognition (ICDAR), Seattle, Washington, pp. 556–560 (2001)
36. Gonzalez, R., Woods, R.: Digital image processing. Addison Wessley Publishing Company (1992)
37. Grube, W.: Zur Feststellung der Schrifturheberschaft bei Unterschriften mit Hilfe densitrometrischer Verfahren. Master's thesis, Humboldt-Universität Berlin (in German) (1977)
38. Grube, W.: Densitrometrische Schreibdruckanalyse - Ein Mittel im Identifizierungsprozeßauf dem Gebiet der kriminalistischen Handschriften- und Dokumentenuntersuchung. Forum der Kriminalistik 25, 41–44 (1989) (in German)
39. Otsu, N.: A threshold selection method from gray-scale histogram. IEEE Transactions on Systems, Man and Cybernetics (SMC) 8, 62–66 (1978)
40. Cha, S.: Use of Distance Measures in Handwriting Analysis. PhD thesis, State University of New York, Buffalo (2001)
41. Srihari, S., Cha, S., Arora, H., Lee, S.: Individuality of handwriting. Journal of Forensic Sciences 47(4), 856–872 (2002)
42. Franke, K., Grube, G.: The automatic extraction of pseudodynamic information from static images of handwriting based on marked grayvalue segmentation (extended version). Journal of Forensic Document Examination 11(3), 17–38 (1998)
43. Franke, K.: The Influence of Physical and Biomechanical Processes on the Ink Trace - Methodological foundations for the forensic analysis of signatures. PhD thesis, Artifical Instelligence Institute, University of Groningen, The Netherlands (2005)
44. Sabourin, R., Genest, G., Preteux, F.: Off-line signature verification by local granulometric size distributions. IEEE Transactions on Pattern Analysis and Machine Intelligence (PAMI) 19(9), 976–988 (1997)
45. Franke, K., Schomaker, L., Penk, W.: On-line pen input and procedures for computer-assisted forensic handwriting examination. In: Teulings, H., van Gemmert, A. (eds.) Proc. 11th Conference of the International Graphonomics Society (IGS), Scottsdale, Arizona, USA, pp. 295–298 (2003)
46. Zimmer, A., Ling, L.: A hybrid on/off line handwritten signature verification system. In: Proc. International Conference on Document Analysis and Recognition (ICDAR), Edinburgh, Scotland (2003)
47. Doermann, D.: Document Image Understanding: Integration Recovery and Interpretation. PhD thesis, University of Maryland (1993)

Invariants Discretization for Individuality Representation in Handwritten Authorship

Azah Kamilah Muda[1], Siti Mariyam Shamsuddin[1], and Maslina Darus[2]

[1] Faculty of Computer Science and Information Systems
Universiti Teknologi Malaysia, Malaysia
[2] Faculty of Sciences and Technology
Universiti Kebangsaan Malaysia, Malaysia
{azah@utem.edu.my, mariyam@utm.my, maslina}@pkrisc.cc.ukm.my

Abstract. Writer identification is one of the areas in pattern recognition that have created a center of attention by many researchers to work in. Its focal point is in forensics and biometric application as such the writing style can be used as biometric features for authenticating a writer. Handwriting style is a personal to individual and it is implicitly represented by unique features that are hidden in individual's handwriting. These unique features can be used to identify the handwritten authorship accordingly. Many researches have been done to develop algorithms for extracting good features that can reflect the authorship with good performance. However, this paper investigates the individuality representation of individual features through discretization technique. Discretization is a procedure to explore the partition of attributes into intervals and to unify the values for each interval. It illustrates the pattern of data systematically which improved the identification accuracy. An experiment has been conducted using IAM database with 3520 training data and 880 testing data (70% training data and 30% testing data) and 2639 training data and 1760 testing data (60% training data and 40% testing data). The results reveal that with invariants discretization, the accuracy of handwritten identification is improved significantly with the classification accuracy of 99.90% compared to undiscretized data.

Keywords: Writer Identification, Authorship Invarianceness, Invariants Discretization.

1 Introduction

Pattern recognition is imperative in various engineering and scientific disciplines such as computer vision, marketing, biology, psychology, medicine, artificial intelligence, remote sensing and etc. One of the areas in pattern recognition is handwriting analysis. Handwriting analysis is important in forensic application such as Writer Identification (WI). Writer identification (WI) can be considered as a particular kind of dynamic biometric since the shape and style of writing can be used as biometric features for authenticating an identity [1-4], similar to signature, fingerprint, iris or face identification. Frequently, writer identification performed on legal papers by a way of signature. However, there is also exist a scenario where to identify a handwritten

S.N. Srihari and K. Franke (Eds.): IWCF 2008, LNCS 5158, pp. 218–228, 2008.

document without a signature such as in a threaten letter, authorship determination of old or historical manuscript, film script (to identify the original idea) and others. In this work, the shape of cursive word is employed and extracted to obtain the features with a proposed descritized process prior to identification task.

Handwriting is individualistic where consistent individual's features are hidden in the shape and writing style. The writing styles are different from one to another, but it is personal to individual. Any written word by the same author must have the same characteristic features, despite of the word shape or writing style. The main issue in writer identification is to acquire the features that reflect the author for varieties of handwriting [3, 5-9] and more important is the unique individual features of handwriting. Previous works have developed new approach or technique for better feature extraction and to proof the individuality concept in handwriting. However, from the literature we found that most of the works are focus on how to extract the individual features and not on illustrating the individual characteristic of handwriting with systematic representation.

The performance of pattern recognition largely depends on the feature extraction and classification/learning scheme [10 - 11]. These two tasks are vital to achieve a good performance in identifying handwritten authorship. Extracting and selecting the meaningful features are a crucial task in the process of pattern recognition prior to classification task, where the extracted features will be classified into categories. Low performance in terms of accuracy is due to various features are representing the same author. It makes the identification process become intricate and complex. The same characteristics are easily identified if all of different features values for same author are having a standard representation for the generalized unique features or individual features. It can make the identification process simpler. Therefore, illustration of individuality features is required to portray the individual's unique features in a systematic representation. This can be achieved by executing the discretization process to demonstrate the pattern of individual features thoroughly.

This paper focal point is to investigate the invariant discretization process of features in order to represent the individual features of writers and significantly illustrates related features in systematic way. In return, it is easily classified and performed better identification result. The paper is systematized as follows. Individuality of handwriting is explained in Section 2. Followed with the authorship invarianceness of moment in Section 3. Section 4 describes the proposed approach of invariants descritization process in this work. The experiment and results is discussed in Section 5. And finally, the conclusion is drawn in Section 6.

2 Individuality of Handwriting

Handwriting has long been considered individualistic and writer individuality rests on the hypothesis that each individual has consistent handwriting [12 -16]. The relation of character, words, shape or style of writing is different from one person to another. Even for one person, they are different in times. However, there are still unique features for each person. These unique features can be generalized as individual's handwriting even though one person has many styles of writing. Fig. 1 is example of words by different authors. Each person's handwriting is seen as having a specific texture [4]. The shape is slightly different for the same author and quite difference for different authors. It shows that each person has its individual style in handwriting. Intra-class measurement is exhibited for features of the same author, and inter-class for different authors. To benchmark these measurements, similarity error is computed for both inter-class and intra-class where the similarity error for intra-class must be lowers than inter-class. This reflects the individuality concept in handwriting. This is called as authorship invarianceness in this work due to the concept of moment function. Moment function is used to extract the features in this work.

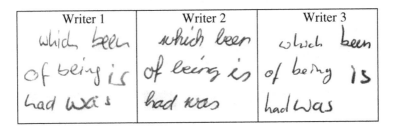

Fig. 1. Various words for different writer

Each image of word is performed the feature extraction task to obtain the features of the image. In this work, the images are extracted to obtain the invariant features using the proposed moment function of integrated Aspect Scale Invariant (ASI) into United Moment Invariant (UMI). It is based on the original United Moment Invariant function. Detail procedure on proposed moment function of integrated ASI into UMI can be referred in [17]. Example of extracted features is shown in Fig 2. Further stage, the extracted features are performed authorship invarianceness analysis to evaluate the individuality concept of handwriting in WI.

3   Authorship Invarianceness

An invarianceness in the context of moment functions can be defined as the *perseverance of the images regardless of its transformations.* In this work, the invarianceness of authorship in WI is given as small similarity error for intra-class (same writer) and large similarity error for inter-class (different writers) of words and regardless of word shape. This is due to the uniqueness features of person in handwriting that called as individuality of handwriting concept in handwriting analysis. The main process of identification in WI is to look for similar characteristic of handwriting based on the nearest unknown handwriting in the database. This can be solved by implementing the individuality of handwriting concept. To achieve this, intra-class and inter-class measurement are implemented to find the nearest characteristic using word shape with the lowest Mean Absolute Error (MAE) value in order to obtain authorship invarianceness. Intra-class should give smaller MAE value compared to inter-class, regardless of any types of word. The range of deviation between intra-class MAE value and inter-class MAE value is not a concern. This is due to the characteristic of Moment

Function where the intra-class value must be lower than inter-class value confirm it can be classified as authorship invarianceness. The MAE function is given by Equation (1):

$$MAE = \frac{1}{n}\sum_{i=1}^{n}\left|(x_i - r_i)\right|$$ (1)

where :

n is number of image.
x_i is the current image.
r_i is the reference image.
i is the feature's column of image.

The result in Table 1 and Table 2 show that the proposed technique of ASI into UMI is worth for further exploration in WI domain. The initial result of similarity error shows that invarianceness of authorship for intra-class (same writer) is smaller compared to inter-class (different writers) for same word and different words, respectively. It is proof the individuality handwriting concept in WI, where MAE value for intra-class (same writer) is smaller value compared to inter-class (different writers) for the same or different words, regardless of short or long word such as the word of "To" or "Being". This is due to the capability of Moment Function in extracting object shape without any constrain in terms of length. Thus, this authorship invarianceness analysis confirms the integrated ASI into UMI techniques can be used to extract features for WI domain.

Table 1. Invarianceness of Authorship for Same Word

Word	Intra-class (1 writer)	Inter-class (10 writers)	Inter-class (30 writers)	Inter-class (60 writers)
To	1.08086	1.10181	1.21423	1.2927
He	0.486922	0.865588	0.721937	0.737597
Of	0.486201	0.702867	0.691087	0.754485
Is	0.489428	0.599104	0.684848	0.779217
Had	0.454727	0.566663	0.670911	0.675404
And	0.564578	0.856195	0.797005	0.782162
The	0.39991	0.718456	0.643291	0.611504
Was	0.736664	0.951713	1.0253	0.955763
Been	1.02514	1.35783	1.28346	1.27161
That	0.677631	1.0147	0.847687	0.768499
With	0.394996	0.706262	0.739905	0.718119
Which	0.335732	0.491985	0.556506	0.599928
Being	0.291463	0.557977	0.581267	0.552889

Table 2. Invarianceness of Authorship for Various Word

Various words	Intra-class (1 writer)	Inter-class (10 writers)	Inter-class (30 writers)	Inter-class (60 writers)
60 words	0.733659	1.11315	0.931423	0.882049
90 words	0.693564	1.03028	0.94499	0.924337
120 words	0.852839	0.975387	0.939999	0.936329

The uniqueness or individual characteristic for each writer in handwriting describes the above result. Similarity error for inter-class (different writers) should be higher than intra-class (same writer) in authorship invarianceness concept. It has been proven in Table 1 and Table 2. For further exploration, these similarity errors can be associated into discretization technique in order to illustrate the data by discerning the individual features into category. The idea is to acquire objects, attributes, decision values, and generate rules for lower, upper and boundary approximations of the set. With these rules, a new object can easily be classified into one of the region or interval which is called as discretization process.

4   Discretization

Discretization is a process of dividing the range of continuous attributes into disjoint regions (interval) which labels can then be used to replace the actual data values [18]. It engages searching for "cuts" that determine intervals and unifying the values over each interval. All values that lie within each interval are mapped to the same value, in effect converting numerical attributes that can be treated as being symbolic [19]. Empirical results show the superiority of classification methods depends on the discretization algorithm used in preprocessing process. There are abundant of discretization algorithms exist based on three basic perspectives. They are supervised versus unsupervised, global versus local and dynamic versus static [20]. Supervised method considers class information is on hand and no classification information available for unsupervised. Another perspective of global versus local describes global method discritized entire data before classification while local method discretized specific amount of defined data. Furthermore, static versus dynamic perspective explains static method discretized each attribute independently without consider interaction between attributes. Meanwhile, dynamic method considered attributes interdependencies while discretization process.

Proposed discretization method is resemblance with the simplest unsupervised methods of Equal Width Binning. However, proposed method is categorized in supervised method because it needs class information to perform discretization process. It globally process for all integrated invariants feature vector for all writers with dynamic characteristic of features in WI domain. The continuous values of invariant features are discretized to obtain the detachment of authors' individuality for better data representation. In this work, invariant features are in real value format, extracted using integrated ASI into UMI technique. Discretization of real value attributes is an essential task in data mining, predominantly the classification problem. Our results disclose that the performance of the classification on writers' handwriting is much improved with discretized data of proposed Invariant Discretization algorithm.

4.1 Proposed Invariant Discretization Algorithm

Discretization is important in this work because it leads to the better accuracy in classification phase compared to undiscretize data. Proposed discretization algorithm is applied where class information is given for the each image to represent the writer. In the process of discretization, it will search the suitable set of cuts to represent the real data for each writer. It divides the range of minimum to maximum data of each writer with the equal size of interval or cuts. Lower and upper approximation is given to the each cut. Number of cuts is defined based on number of feature vector for the each word image, i.e, eight feature vector values of ASI into UMI are used to represent a pattern of image. This is to keep the original number of invariant vector in moment invariant function that has been applied. Each cuts will represented with one defined representation value. Feature's values that fall within the same cut will have the same representation. The proposed discretization algorithm is given below :

Algorithm of Proposed Discretization

```
For each writer {
   Min = min feature;  Max = max feature;
   No_bin = no_feature_invariant;
   Interval = (Max - Min)/ No_bin;

   For each bin {
      Find lower and upper value of interval;
      RepValue = (upper -lower)/2;
   }

   For (1 to no_feature_invariant) {
      For each bin {
         If (feature in range of interval)
            Dis_Feature = RepValue;
      }
   }
}
```

Process to calculate the interval and representation value for the each cut is done based on writer classes. This is due to the concept in WI domain where each person has their own style of writing or individuality in handwriting. To make sure the uniqueness or individuality characteristic is preserved, the interval and representation value is calculated based on each writer. If there are two different writers that have closed or same invariant feature, there will be the same or quite similar interval or cuts for these two classes. Therefore, the representation value of each cut will be same or quite similar. Thus, this proposed algorithm is not changed the information gather or characteristic of writers. It just represents the real invariant data into better data representation. Discretization process is implemented to illustrate the features clearly and not to change the characteristics of features. Therefore, the proposed discretization of each writer's class approach is seen as acceptable and match with the individuality concept in WI.

Example of transformation of feature invariant vector to discretize feature vector is ilustrated in Fig. 2 through Fig. 4 below. Fig. 2 shows the example of data before discretization process for various images of writers. There are eight columns of extracted invariant feature vectors and the last column is the label of author's class. Eight invariant vectors of feature in one row represent one word image for the writer in the last column.

2.59224	3.23024	0.332166	0.672428	0.617473	4.56811	2.55781	1.02415	1
3.61109	3.62337	0.0471209	0.10731	3.39726	3.82502	3. 51606	0.366274	4
2.91782	3.11856	0.0524496	0.204262	2.40792	3.42825	2.9143	0.43011	1
3.34655	3.40755	0.284003	1.13843	1.57912	5.11418	2.26912	1.43088	1
2.74886	2.75738	0.0650583	0.31401	2.29621	3.20228	2.44336	0.512494	2
3.18126	3.18186	0.229476	0.475357	2.24635	4.11646	2.7065	0.752626	8
3.54961	3.74973	0.180705	1.65463	1.33345	5.76589	2.0951	1.8395	8
3.05499	4.58202	0.163657	0.588422	0.612222	5.49814	3.9936	0.801845	2
3.18019	3.49778	0.0694599	0.81009	1.9136	4.44707	2.68769	0.925833	2
3.36354	3.67488	0.115037	0.471541	2.81074	3.91654	3.20334	0.553371	2
3.24221	3.526	0.0506261	0.928334	2.13133	4.35333	2.59767	0.937993	3
3.39974	3.40077	0.0320461	0.249581	3.08504	3.71462	3.15119	0.453545	4
3.50443	5.83822	0.0726602	0.182035	0.843275	6.16572	5.65619	0.422188	4
4.19887	5.14676	0.31637	0.30295	4.18666	4.2111	5.44971	0.364019	6
3.51602	3.57551	0.0456017	0.187287	3.36045	3.67173	3.38822	0.397325	7
2.68472	2.68664	0.357513	0.104451	3.29337	2.07696	2.58219	0.452733	6
3.66434	3.77518	0.176983	0.50243	2.6971	4.63167	3.27275	0.736009	8
3.58092	4.42651	0.139926	0.501636	2.51353	4.64841	3.92488	0.565854	8
3.55531	3.84184	0.177988	0.32283	3.94758	3.16315	4.16467	0.260595	8

Fig. 2. Real data of invariant feature vector

Data in Fig. 2 is continued to perform discretization process as shown in Fig. 3. It is an example to discretize data for writer 1. Discretized feature data of discretization process is shown in Fig. 4 for all data in Fig. 2.

From the discretized feature data in Fig. 4, it shows that each writer has its own representation data which illustrates the characteristic of each writer. It represents the individuality concept of handwriting in WI domain where each person has its own style of handwriting. These discretized features data then undergo identification process in order to analyze the performance of identification.

Discretization for Writer 1 :
Min Value :   0.0524496   Max Value : 5.11418

LOW and UPPER value for BIN for Writer : 1
Bin 0: Low :0.0524496   Upper :0.685166 Rep Value for Bin 0: 0.316358
Bin 1: Low :0.685166   Upper :1.31788 Rep Value for Bin 1: 1.00152
Bin 2: Low :1.31788   Upper :1.9506 Rep Value for Bin 2: 1.63424
Bin 3: Low :1.9506   Upper :2.58331 Rep Value for Bin 3: 2.26696
Bin 4: Low :2.58331   Upper :3.21603 Rep Value for Bin 4: 2.89967
Bin 5: Low :3.21603   Upper :3.84875 Rep Value for Bin 5: 3.53239
Bin 6: Low :3.84875   Upper :4.48146 Rep Value for Bin 6: 4.16511
Bin 7: Low :4.48146   Upper :5.11418 Rep Value for Bin 7: 4.79782

DISCRETIZE DATA
2.89967 3.53239 0.316358 0.316358 0.316358 4.79782 2.26696 1.00152 1
2.89967 2.89967 0.316358 0.316358 2.26696 3.53239 2.89967 0.316358 1
3.53239 3.53239 0.316358 1.00152 1.63424 4.79782 2.26696 1.63424 1

Fig. 3. Example of Discretization Process for Writer 1

2.89967	3.53239	0.316358	0.316358	0.316358	4.79782	2.26696	1.00152	1
2.89967	2.89967	0.316358	0.316358	2.26696	3.53239	2.89967	0.316358	1
3.53239	3.53239	0.316358	1.00152	1.63424	4.79782	2.26696	1.63424	1
3.48224	3.48224	0.383355	0.383355	3.48224	3.48224	3.48224	0.383355	4
3.48224	3.48224	0.383355	0.383355	2.71553	3.48224	3.48224	0.383355	4
3.48224	5.78237	0.383355	0.383355	1.18211	5.78237	5.78237	0.383355	4
2.44203	2.44203	0.339568	0.339568	2.44203	3.12117	2.44203	0.339568	2
3.12117	4.47944	0.339568	0.339568	0.339568	5.15857	3.8003	1.08376	2
3.12117	3.8003	0.339568	1.08376	1.7629	4.47944	2.44203	1.08376	2
3.12117	3.8003	0.339568	0.339568	3.12117	3.8003	3.12117	0.339568	2
3.30453	3.30453	0.351623	0.351623	1.89804	4.00778	2.60129	0.351623	8
3.30453	4.00778	0.351623	1.89804	1.19479	5.41427	1.89804	1.89804	8
4.00778	4.00778	0.351623	0.351623	2.60129	4.71102	3.30453	0.351623	8
3.30453	4.71102	0.351623	0.351623	2.60129	4.71102	4.00778	0.351623	8
3.30453	4.00778	0.351623	0.351623	4.00778	3.30453	4.00778	0.351623	8
3.00874	3.54657	0.268919	0.857383	1.93306	4.08441	2.4709	0.857383	3
4.44747	5.11563	0.334079	0.334079	4.44747	4.44747	5.11563	0.334079	6
2.443	2.443	0.334079	0.334079	3.11116	1.77484	2.443	0.334079	6
3.4451	3.4451	0.226633	0.226633	3.4451	3.4451	3.4451	0.226633	7

Fig. 4. Example of Descritized Feature Data

5   Experiment Result and Discussion

Experiment is conducted to proof the discretization process can improve the performance of identification in WI domain. The comparisons of identification accuracy (%) for discretized data with un-discretize data are shown in Table 3 and

Table 4. Two techniques have been used to extract the features from the various written words, which are original UMI and proposed ASI into UMI. Identification accuracy is compared for these two techniques. For identification task, discretized data and un-discretized data are run with Johnson Algorithm and 1R Algorithm, which are the techniques that embedded in Rosetta toolkit [21]. Meanwhile R-Chunk is the pattern matching that applied in Modified Negative Selection Algorithm (MNSA) classifier [22]. Un-discretized data is the original extracted features meanwhile discretized data is the extracted features that performed discretized process using proposed invariant discretization algorithm. IAM database [23] with 60 writers from the various types of word images were used to run this experiment. Table 3 is for 3520 training data and 880 testing data (70% training data and 30% testing data) and Table 4 is for 2639 training data and 1760 testing data (60% training data and 40% testing data).

Table 3. Comparison of Identification Accuracy for 3520 Training Data and 880 Testing Data

Technique	Original UMI	ASI into UMI	Data
Johnson Algorithm	33.56	35.34	Un_Dis
	99.09	99.55	Dis
1R Algorithma	33.67	35.34	Un_Dis
	99.90	99.90	Dis
R-Chunck Algorithm	45.80	46.68	Un_Dis
	95.34	99.88	Dis

Table 4. Comparison of Identification Accuracy for 2639 Training Data and 1760 Testing Data

Technique	Original UMI	ASI into UMI	Data
Johnson Algorithm	29.92	31.70	Un_Dis
	97.95	98.75	Dis
1R Algorithma	30.03	31.70	Un_Dis
	99.90	99.90	Dis
R-Chunck Algorithm	37.54	38.63	Un_Dis
	95.52	99.89	Dis

Discretized data gives higher accuracy for both feature extraction techniques and all classifiers tested in the experiment. Discretization is performed to represent the features of data systematically. Thus, the individuality of handwriting is clearly illustrated in discretized data. The same characteristics are easily identified if all of different features values for each author are having a standard represented value for the generalized unique features or individual features. Therefore increased the performance of identification compared to un-discretized data. The focus in this paper is to show that discretized data performed much better in identifying author compared to un-discretized data. Both tables show that discretized data give much better performance in identification and it is proven in the experiment result.

6 Conclusion

This paper proposed an approach of invariants discretization to represent the individual features systematically. Discrete features extracted from the various words undergo discretization process for granular mining of writer authorship. Similarity errors are reduced between these data, thus handwritten authorship can be defined easily. It is experimentally evaluated that discretized data give much better performance of identification compared to un-discretized data in WI domain. Our experiments have revealed better results with various identification techniques in classification process of Rosetta toolkit [21] and MNSA classifier[22].

Acknowledgments

This work was supported by Ministry of Higher of Education (MOHE) under Fundamental Research Grant Scheme (Vot 78182). Authors would like to thank Research Management Center (RMC), Universiti Teknologi Malaysia for the research activities and *Soft Computing Research Group* (SCRG) for their support in making this research success.

References

[1] Srihari, S.N., Huang, C., Srinivasan, H., Shah, V.A.: Biometric and forensic aspects of digital document processing. In: Chaudhuri, B.B. (ed.) Digital Document Processing. Springer, Heidelberg (2006)

[2] Tapiador, M., Sigüenza, J.A.: Writer identification method based on forensic knowledge, Biometric Authentication. In: Zhang, D., Jain, A.K. (eds.) ICBA 2004. LNCS, vol. 3072, pp. 555–561. Springer, Heidelberg (2004)

[3] Kun, Y., Yunhong, W., Tieniu, T.: Writer identification using dynamic features, Biometric Authentication. In: Zhang, D., Jain, A.K. (eds.) ICBA 2004. LNCS, vol. 3072, pp. 512–518. Springer, Heidelberg (2004)

[4] Zhu, Y., Tan, T., Wang, Y.: Biometric personal identification based on handwriting, Pattern Recognition. In: 15th International Conference, September 3-7, vol. 2, pp. 797–800 (2000)

[5] Bensefia, A., Paquet, T., Heutte, L.: A writer identification and verification system. Pattern Recognition Letters 26(10), 2080–2092 (2005)

[6] Schlapbach, A., Bunke, H.: Off-line handwriting identification using HMM based recognizers. In: 17th Int. Conf. on Pattern Recognition, Cambridge, UK, August 23-26, pp. 654–658 (2004)

[7] He, Z.Y., Tang, Y.Y.: Chinese handwriting-based writer identification by texture analysis, Machine Learning and Cybernetics. In: 2004 International Conference, August 26-29, vol. 6, pp. 3488–3491 (2004)

[8] Srihari, S.N., Cha, S.-H., Arora, H., Lee, S.: Individuality of handwriting. Journal of Forensic Sciences 47(4), 1–17 (2002)

[9] Shen, C., Ruan, X.-G., Mao, T.-L.: Writer identification using Gabor wavelet, Intelligent Control and Automation. In: 4th World Congress, June 10-14, vol. 3, pp. 2061–2064 (2002)

[10] Liu, C.-L., Nakashima, K., Sako, H., Fujisawa, H.: Handwritten digit recognition: benchmarking of state-of-the-art techniques. Pattern Recognition 36(10), 2271–2285 (2003)

[11] Liu, C.-L., Nakashima, K., Sako, H., Fujisawa, H.: Handwritten digit recognition: investiga-tion of normalization and feature extraction techniques. Pattern Recognition 37(2), 265–279 (2004)

[12] Srihari, S.N., Huang, C., Srinivasan, H., Shah, V.A.: Biometric and forensic aspects of digital document processing. In: Chaudhuri, B.B. (ed.) Digital Document Processing. Springer, Heidelberg (2006)

[13] Zhang, B., Srihari, S.N.: Analysis of handwriting individuality using word features, Document Analysis and Recognition. In: Seventh International Conference, August 3-6, pp. 1142–1146 (2003)

[14] Srihari, S.N., Cha, S.-H., Arora, H., Lee, S.: Individuality of handwriting. Journal of Forensic Sciences 47(4), 1–17 (2002)

[15] Srihari, S.N., Cha, S.-H., Lee, S.: Establishing handwriting individuality using pattern recog-nition techniques, Document Analysis and Recognition. In: Sixth International Conference on Document Analysis and Recognition (ICDAR 2001), Seattle, September 10-13, pp. 1195–1204 (2001)

[16] Zhu, Y., Tan, T., Wang, Y.: Biometric personal identification based on handwriting, Pattern Recognition. In: 15th International Conference, September 3-7, vol. 2, pp. 797–800 (2000)

[17] Muda, A.K., Shamsuddin, S.M., Darus, M.: Embedded scale united moment invariant for identification of handwriting individuality. In: Gervasi, O., Gavrilova, M.L. (eds.) ICCSA 2007, Part I. LNCS, vol. 4705, pp. 385–396. Springer, Heidelberg (2007)

[18] Agre, G., Peev, S.: On supervised and unsupervised discretization. Cybernetics And Information Technologies 2(2), 43–57 (2002)

[19] Nguyen, H.S.: Discretization problems for rough set methods. In: Polkowski, L., Skowron, A. (eds.) RSCTC 1998. LNCS (LNAI), vol. 1424, pp. 545–552. Springer, Heidelberg (1998)

[20] Dougherty, J., Kohavi, R., Sahami, M.: Supervised and unsupervised discretization of con-tinuous features. In: Twelfth International Conference on Machine Learning, pp. 194–202. Morgan Kaufmann, Los Altos (1995)

[21] Øhrn, A., Komorowski, J.: ROSETTA: A rough set toolkit for analysis of data. In: Wang, P.P. (ed.) Proc. Third International Joint Conference on Information Sciences, Durham, NC, vol. 3, pp. 403–407 (March 1997)

[22] Muda, A.K., Shamsuddin, S.M., Darus, M.: Bio-inspired generalized global shape approach for writer identification. Transaction on Engineering, Computing and Technology 16, 55–59 (2006)

[23] Marti, U.-V., Bunke, H.: The IAM-database: an english sentence database for off-line handwriting recognition. Int. Journal on Document Analysis and Recognition 5, 39–46 (2002)

Author Index

Lecture Notes in Computer Science

Sublibrary 6: Image Processing, Computer Vision, Pattern Recognition, and Graphics

For information about Vols. 1– 3951
please contact your bookseller or Springer

Vol. 4569: A. Butz, B. Fisher, A. Krüger, P. Olivier, S. Owada (Eds.), Smart Graphics. IX, 237 pages. 2007.

Vol. 4538: F. Escolano, M. Vento (Eds.), Graph-Based Representations in Pattern Recognition. XII, 416 pages. 2007.

Vol. 4522: B.K. Ersbøll, K.S. Pedersen (Eds.), Image Analysis. XVIII, 989 pages. 2007.

Vol. 4485: F. Sgallari, A. Murli, N. Paragios (Eds.), Scale Space and Variational Methods in Computer Vision. XV, 931 pages. 2007.

Vol. 4478: J. Martí, J.M. Benedí, A.M. Mendonça, J. Serrat (Eds.), Pattern Recognition and Image Analysis, Part II. XXVII, 657 pages. 2007.

Vol. 4477: J. Martí, J.M. Benedí, A.M. Mendonça, J. Serrat (Eds.), Pattern Recognition and Image Analysis, Part I. XXVII, 625 pages. 2007.

Vol. 4472: M. Haindl, J. Kittler, F. Roli (Eds.), Multiple Classifier Systems. XI, 524 pages. 2007.

Vol. 4466: F.B. Sachse, G. Seemann (Eds.), Functional Imaging and Modeling of the Heart. XV, 486 pages. 2007.

Vol. 4418: A. Gagalowicz, W. Philips (Eds.), Computer Vision/Computer Graphics Collaboration Techniques. XV, 620 pages. 2007.

Vol. 4417: A. Kerren, A. Ebert, J. Meyer (Eds.), Human-Centered Visualization Environments. XIX, 403 pages. 2007.

Vol. 4391: Y. Stylianou, M. Faundez-Zanuy, A. Esposito (Eds.), Progress in Nonlinear Speech Processing. XII, 269 pages. 2007.

Vol. 4370: P.P. Lévy, B. Le Grand, F. Poulet, M. Soto, L. Darago, L. Toubiana, J.-F. Vibert (Eds.), Pixelization Paradigm. XV, 279 pages. 2007.

Vol. 4358: R. Vidal, A. Heyden, Y. Ma (Eds.), Dynamical Vision. IX, 329 pages. 2007.

Vol. 4338: P.K. Kalra, S. Peleg (Eds.), Computer Vision, Graphics and Image Processing. XV, 965 pages. 2006.

Vol. 4319: L.-W. Chang, W.-N. Lie (Eds.), Advances in Image and Video Technology. XXVI, 1347 pages. 2006.

Vol. 4292: G. Bebis, R. Boyle, B. Parvin, D. Koracin, P. Remagnino, A. Nefian, G. Meenakshisundaram, V. Pascucci, J. Zara, J. Molineros, H. Theisel, T. Malzbender (Eds.), Advances in Visual Computing, Part II. XXXII, 906 pages. 2006.

Vol. 4291: G. Bebis, R. Boyle, B. Parvin, D. Koracin, P. Remagnino, A. Nefian, G. Meenakshisundaram, V. Pascucci, J. Zara, J. Molineros, H. Theisel, T. Malzbender (Eds.), Advances in Visual Computing, Part I. XXXI, 916 pages. 2006.

Vol. 4245: A. Kuba, L.G. Nyúl, K. Palágyi (Eds.), Discrete Geometry for Computer Imagery. XIII, 688 pages. 2006.

Vol. 4241: R.R. Beichel, M. Sonka (Eds.), Computer Vision Approaches to Medical Image Analysis. XI, 262 pages. 2006.

Vol. 4225: J.F. Martínez-Trinidad, J.A. Carrasco Ochoa, J. Kittler (Eds.), Progress in Pattern Recognition, Image Analysis and Applications. XIX, 995 pages. 2006.

Vol. 4191: R. Larsen, M. Nielsen, J. Sporring (Eds.), Medical Image Computing and Computer-Assisted Intervention – MICCAI 2006, Part II. XXXVIII, 981 pages. 2006.

Vol. 4190: R. Larsen, M. Nielsen, J. Sporring (Eds.), Medical Image Computing and Computer-Assisted Intervention – MICCAI 2006, Part I. XXXVVIII, 949 pages. 2006.

Vol. 4179: J. Blanc-Talon, W. Philips, D. Popescu, P. Scheunders (Eds.), Advanced Concepts for Intelligent Vision Systems. XXIV, 1224 pages. 2006.

Vol. 4174: K. Franke, K.-R. Müller, B. Nickolay, R. Schäfer (Eds.), Pattern Recognition. XX, 773 pages. 2006.

Vol. 4170: J. Ponce, M. Hebert, C. Schmid, A. Zisserman (Eds.), Toward Category-Level Object Recognition. XI, 618 pages. 2006.

Vol. 4153: N. Zheng, X. Jiang, X. Lan (Eds.), Advances in Machine Vision, Image Processing, and Pattern Analysis. XIII, 506 pages. 2006.

Vol. 4142: A. Campilho, M.S. Kamel (Eds.), Image Analysis and Recognition, Part II. XXVII, 923 pages. 2006.

Vol. 4141: A. Campilho, M.S. Kamel (Eds.), Image Analysis and Recognition, Part I. XXVIII, 939 pages. 2006.

Vol. 4122: R. Stiefelhagen, J.S. Garofolo (Eds.), Multimodal Technologies for Perception of Humans. XII, 360 pages. 2007.

Vol. 4109: D.-Y. Yeung, J.T. Kwok, A. Fred, F. Roli, D. de Ridder (Eds.), Structural, Syntactic, and Statistical Pattern Recognition. XXI, 939 pages. 2006.

Vol. 4091: G.-Z. Yang, T. Jiang, D. Shen, L. Gu, J. Yang (Eds.), Medical Imaging and Augmented Reality. XIII, 399 pages. 2006.

Vol. 4073: A. Butz, B. Fisher, A. Krüger, P. Olivier (Eds.), Smart Graphics. XI, 263 pages. 2006.

Vol. 4069: F.J. Perales, R.B. Fisher (Eds.), Articulated Motion and Deformable Objects. XV, 526 pages. 2006.

Vol. 4057: J.P.W. Pluim, B. Likar, F.A. Gerritsen (Eds.), Biomedical Image Registration. XII, 324 pages. 2006.

Vol. 4046: S.M. Astley, M. Brady, C. Rose, R. Zwiggelaar (Eds.), Digital Mammography. XVI, 654 pages. 2006.

Vol. 4040: R. Reulke, U. Eckardt, B. Flach, U. Knauer, K. Polthier (Eds.), Combinatorial Image Analysis. XII, 482 pages. 2006.

Vol. 4035: T. Nishita, Q. Peng, H.-P. Seidel (Eds.), Advances in Computer Graphics. XX, 771 pages. 2006.

Vol. 3979: T.S. Huang, N. Sebe, M. Lew, V. Pavlović, M. Kölsch, A. Galata, B. Kisačanin (Eds.), Computer Vision in Human-Computer Interaction. XII, 121 pages. 2006.

Vol. 3954: A. Leonardis, H. Bischof, A. Pinz (Eds.), Computer Vision – ECCV 2006, Part IV. XVII, 613 pages. 2006.

Vol. 3953: A. Leonardis, H. Bischof, A. Pinz (Eds.), Computer Vision – ECCV 2006, Part III. XVII, 649 pages. 2006.

Vol. 3952: A. Leonardis, H. Bischof, A. Pinz (Eds.), Computer Vision – ECCV 2006, Part II. XVII, 661 pages. 2006.